8/10/16

exploding stars

dark energy

and the
accelerating cosmos

the extravagant universe

Robert P. Kirshner

With a new epilogue by the author

Princeton University Press
Princeton and Oxford

GB
843
.S95
K57
2016

Fourth printing, and first paperback printing, for the Princeton Science Library,
with a new epilogue, 2004
New Princeton Science Library edition, 2016
Paperback ISBN 978-0-691-17318-4

The cloth edition of this book has been cataloged as follows:

Library of Congress Control Number 2002029268
ISBN 0-691-05862-8 (cloth : alk. paper)

British Library Cataloging-in-Publication Data is available

This book has been composed in ITC Garamond Light

Printed on acid-free paper. ∞

press.princeton.edu

Printed in the United States of America

10 9 8 7 6 5 4 3 2 1

For Rebecca and Matthew—

who will live in an ever-larger universe

contents

preface

Since 1970, I have been among the astronomers observing the exploding stars known as supernovae to learn what they are, how they work, and how they affect the chemistry of the universe. As a bonus, this investigation created ways to convert supernovae into the best cosmic yardsticks for measuring distances in the universe. One variety of supernova comes from the thermonuclear explosion of a dense stellar clinker left over from a burnt-out star like the sun. These "type Ia supernovae" (SN Ia) make useful standard bombs whose distances can be accurately judged from their apparent brightness. Using SN Ia as a sailor might use a lighthouse to judge distances at sea, we can measure the distances to galaxies, the giant pinwheels and bloated zeppelins of stars in which supernovae explode.[1] Remarkably, measuring the distances to supernovae has led to a dramatic new picture for the contents of the universe, dominated by a dark energy that springs from the properties of empty space itself.

Since 1912, astronomers have measured the motions of galaxies. Almost every one is moving away from our own Milky Way galaxy, the phenomenon known as the redshift. In 1929, Edwin Hubble connected the distances to galaxies with their redshifts, showing that distant galaxies recede more rapidly than our neighbors. This means we live in an expanding universe.

News of the expanding universe came as a big surprise to Albert Einstein. Back in 1917, when he consulted astronomers they had told him the universe was static. His newly invented theory of general relativity predicted either an expanding universe or a contracting one. But you can't fight the facts, even when they are wrong. Einstein sighed and stuck in a mathematical constant to fix this "problem" by inventing an expansive quality of space itself, which today we call "dark energy," to balance the inward pull of gravitating matter. Einstein's term, the cosmological constant, was introduced to make the universe stand still, balanced like a skilled cyclist at a stoplight. When, a decade later, Einstein learned that Hubble's new astronomical observations showed that the universe was *not* standing still, he wasn't slow to throw the cosmological constant overboard. "It was theoretically unsatisfactory anyway," he said.[2] The cosmological constant was banished from most serious discussions of cosmology. Who needed it?

By 1990, as astronomers slowly constructed an inventory for the contents of the universe, we ran into a problem, a puzzle, and a conundrum. The problem is that most of the gravitating material in the universe is invisible, the puzzle is that there is not enough of it, and the conundrum is that having enough of this dark matter would have the bad side effect of making the universe younger than its contents. Being invisible is not so bad—we can detect the *effects* of invisible mass even if it emits no light, just as a sailor knows an invisible puff of wind is coming by watching the riffles it makes on the water. Visible matter drains into the invisible web of cosmic troughs that cold dark matter forms. But the puzzle remains that the amount of matter in the universe is only about one-third of the amount that our favorite theories required to make the neatest universe. What's worse is the conundrum posed by cosmic timescales. The oldest stars in our galaxy appear to be about 12 billion years old. If the universe had its full load of gravitating matter, cosmic expansion should slow over time, and the universe would have clocked an elapsed time since the Big Bang of about 10 billion years. Having 12 billion year old stars in a 10 billion year old universe doesn't inspire confidence that this is a genuine history of the physical world. What's wrong with this picture? Are these small cracks in a beautiful fresco, do they show we have a serious conceptual problem with the Big Bang, or is something missing?

In the past several years, teams of scientists have been using new instruments and new telescopes, including the Hubble Space Telescope, to find distant supernovae. These let you measure directly the history of cosmic expansion. We expected to see how much the universe has been slowing down since the Big Bang. I have been involved with one of these teams, a cheerful, slightly anarchic band of brothers (with some sisters, too) that we call the high-z supernova search team. The letter "z" is the astronomer's shorthand for redshift, so this means we've been looking for exploding stars at large redshifts and large distances.

In 1997, this work was well underway when I was invited to Princeton University to give a series of lectures that became the foundation for this book. But, looking over my old notes, I see that we had almost no results to report in 1997: though we knew what the questions were, and saw how to get the answers, the surprising solution to these astronomical riddles has come together in a rush since then. So I talked a lot about how supernovae explode and make new chemical elements and only a little about the way that supernovae would measure the history of cosmic expansion. Now the preliminary results are in and we have a new and surprising synthesis that solves the problems, puzzles, and conundrums of a decade ago.

The observations of distant supernovae show that we live in a universe that is not static as Einstein thought, and not just expanding as Hubble showed, but accelerating! We attribute this increase in expansion over time to a dark energy with a strange type of pressure. In its simplest form this might be Einstein's cosmological constant, which for 60 years was theoretical poison ivy. Dark energy makes up the missing component of mass–energy that theorists have sought, reconciles the ages of objects with the present expansion rate of the universe, and complements new measurements of the lingering glow of the Big Bang itself to make a neat and surprising picture for the contents of the universe.

The last five years have been a little like that moment in assembling a jigsaw puzzle when you complete the frame, pieces are dropping rapidly into place, and you can even see the shapes of the missing pieces. The missing piece may be the most important. A universe controlled by dark energy points to a deep gap in our understanding of submicroscopic aspects of empty space: the properties of the vacuum. No laboratory experiment measures and no physical theory predicts the amount of dark energy our observations imply. The next step forward in understanding the universe on the smallest scale will be to meld gravity with the other forces of nature. Perhaps when there is a new theoretical vision this extravagant universe, propelled by dark energy, will seem simple and inevitable. But for the moment, solving the mysteries of the accelerating universe has produced another delicious puzzle to investigate.

Our working picture of the universe today is extravagant: it has neutrinos as hot dark matter; something unknown as cold dark matter; inflation in the first 10^{-35} second after the Big Bang; and acceleration by dark energy now, when the universe is 10^{52} times older. This is wilder than anyone imagined, but it is based on evidence even though all of these things are invisible. We've built this picture by observing light from the Big Bang itself; from stars, steady, variable, and exploding; and from galaxies at the edge of the observable universe.

Seeing new aspects of the universe for the very first time is a pleasure experienced by the hard-working people who appear in this book. But why should we have all the fun? My aim is to help you share in this adventure where the thrill comes from understanding.

the
extravagant
universe

the big picture

At first, the idea of understanding the universe seems preposterous, presumptuous, or in any case, out of reach, precisely because the universe is not built on a human scale of time or size. But we now have a physical picture of the history and evolution of the universe. How have we overcome the limitations of our small brains, our short lives, and our absurdly small stature to understand an ancient and immense universe?

We're so *brief*. The stars seem permanent, but that's only because we're just passing through. If you live for 100 years that's only one part in 100 million of the age of the universe. How can you expect to see the flow of cosmic change? Comparing your lifetime to the age of the universe is like comparing the longest time you can hold your breath to your lifetime. That's it. One breath is to one lifetime as one lifetime is to the age of universe. Inhale deeply!

Cosmic time numbs our sense of history. All of recorded human history reaches back only 10,000 years: 100 generations for 100 years each. Deep cosmic time stretches back a *million* times farther than the first glimmer of civilization when dogs decided to join humans in their caves. With a few spectacular exceptions, as when stars destroy themselves in supernova explosions, we have no chance to see the universe change during one lifetime, even though we know processes of change must be at work. But by learning what supernovae are, how they work, and how to use them, we can trace the history of cosmic expansion deep into the distant past.

And we're *short*. So short that we can't see the curve of the spherical Earth, which is 10 million times bigger than a person. Our common sense view of a flat Earth is wrong because the Earth, to say nothing of larger astronomical objects, is not built to our scale.[1] We usually learn our planet's shape by meekly accepting dogma from third-grade teachers teaching the Columbus Day curriculum. A better way is to launch people off Earth's surface to take a look. Astronauts travel for us and bring back pictures that illuminate the true spherical geometry of the Earth. Even though we knew what these pictures would show, images of a round planet conquer our common sense and move a spherical Earth into our intuition.

Stepping back to get perspective doesn't work so well for learning the shape of larger astronomical objects. Just as a slice of pepperoni sizzling amid the mozzarella has a hard time seeing the whole pizza, we have a hard time seeing the flattened disk of the galaxy in which the sun is located. We have no perspective on the shape of our Milky Way galaxy and there's no stepping back. Our difficulty in imagining the shape of the universe in which the Milky Way and 100 billion equivalent systems reside is even more acute: there is no way to get outside for some perspective.

How do we overcome these limitations to gain a picture of the universe? Although we have small brains, brief lives, and a common sense that seems certain to lead us astray, the case is not altogether desperate. The problem isn't the size of our brains, it's having the right ideas. Over the past 500 years we have begun to puzzle out where we are and how things work.

Human imagination can begin to explore the possibilities. The old German 10-mark note, now displaced by the Euro, depicted Karl Friedrich Gauss, prince of mathematicians. His civil service job was to direct the astronomical observatory at Göttingen. Astronomers invoke his name daily, using his bell-shaped curve to evaluate the effects of chance on every type of astronomical evidence from motions in the solar system all the way out to tracing the bubbling variations in the glow from the hot Big Bang.[2]

Figure 1.1. **The 4-meter Victor and Betty Blanco telescope at Cerro Tololo in Chile, silhouetted against the Milky Way Galaxy.** In 1917, when Einstein first considered the effects of gravity on the universe as a whole, astronomers thought that the Milky Way was the entire universe. Today we think of it as one galaxy among 100 billion similar systems. The Large and Small Magellanic Clouds are to the left. Courtesy of Roger Smith/NOAO/AURA/NSF.

Ideas of curved space were worked out by Gauss in the 1820s and advanced in the 1850s by his brilliant student and colleague at Göttingen, Bernhard Riemann. Being a mathematician, Riemann was not constricted to thinking about two-dimensional spaces like the surface of a beach ball, but thought through general properties of curvature for mathematical spaces with three or four or many more dimensions.

In 1915, Albert Einstein needed those ideas of curved space to construct a new theory of gravity. In Einstein's general relativity, the presence of matter and energy warps a four-dimensional space–time and affects the way light travels through the universe. Mathematics developed by mathematicians for their own reasons turned

AS8209087D3

Figure 1.2. **Karl Friedrich Gauss on the 10-mark note**. Gauss had early success in predicting orbits and became director of the observatory at Göttingen. The bell-shaped curve of probability looming over Gauss's shoulder describes the likelihood of obtaining, by chance, an experimental result that differs from the true value. When astronomers quote the age of the universe with a band of uncertainty, or the odds that the data imply a cosmological constant, they use the ideas of Gauss.

out to be just the tool that Einstein needed to describe the physical world. Gravity is weak here and the solar system is very small, so curved space makes only subtle differences in the solar system, just as the curvature of the Earth makes only subtle differences in laying out a baseball diamond. But over cosmic distances the curvature of space matters. Einstein's general theory of relativity describes the way matter and energy curve the universe and how the contents of the universe make it expand or contract on the biggest imaginable scale. Using exploding stars, the heat left over from the Big Bang, and a strong web of physical understanding developed over centuries, we now have our first real glimpse of cosmic history and cosmic geometry.

No person has to construct our picture of the universe singlehandedly: science lets us accumulate the understanding of very fine brains of the past such as those of Gauss and Einstein, cooperate and compete with other people today, and harness rapidly improving technology to sift vast haystacks of data. Other aspects of culture may or may not have improved from the time of Shakespeare or Mozart or Rembrandt, but science today is most definitely better than the science of past centuries, or even the past decade. We get to use every good idea and measurement from the past because scientists publish their findings in carefully screened journals. We get to use sharp new tools like the Hubble Space Telescope (HST), giant electronic cameras, and powerful computers for present-day exploration. In this way, more-or-less ordinary people today can make far better measurements than Galileo or Newton or Hubble ever could. Since we get to peek at Einstein's homework and have new and powerful tools of observation, we would be dull astronomers indeed if we couldn't make some progress in learning the history of the universe.

We can decode the universe because the laws of physics discovered on Earth also work in distant places. Gravity accelerating a roller coaster (and its thrilled riders) on the Boardwalk at Santa Cruz is just the local form of universal gravitation that keeps planets and asteroids in their orbits, steers stars around in clusters and galaxies, and determines whether the universe will expand forever. Atoms of calcium, whether in your femur, the sun's atmosphere, or in the atmospheres of stars in a distant galaxy, are interchangeable units governed by electrical forces that interact through precisely the same quantum mechanical laws here and there. The way an atom emits or absorbs light in a fluorescent tube in the humming control room of a telescope is identical to the way a similar atom behaves in an exploding star. You can tell which chemical elements are in a star and how that star is moving by gathering its light with a telescope, then delicately dissecting it into a spectrum. Less familiar laws of physics, discovered in particle accelerators on Earth, govern the weak and strong forces that tell how subatomic particles are assembled and how they push and pull on each other. These laws of physics, combined with human imagination and guided by astronomical observations, tell us how the stars shine and what makes some of them explode as supernovae, and let us interpret the clues to the past that a hot, expanding universe leaves behind as evidence.

Despite these successes, human imagination is a weak thing. The universe is wilder than we imagine: we keep underestimating how weird it really is. So astronomy is not exactly an experimental science in which the thoughtful predictions of physical theory get tested. Astronomy is a science driven by discovery, since the objects we observe are stranger and more exotic than even the most unbridled speculators predict. Where the physical effects are simple, astronomy resembles physics. For example, glowing embers of a vanished hot Big Bang can be detected in every direction as a faint radio hiss we call the cosmic microwave background. Predictions and measurements of this background radiation provide sharp tests for the simple physics of a hot Big Bang. But, where the phenomena have many too many moving parts for a simple analysis, astronomical observations lead the way. Once the universe got complex, as matter formed into stars, it grew less predictable and far more interesting. The exact mechanisms by which stars explode in thermonuclear blasts are still not fully understood and were not predicted by even the most uninhibited minds. Yet we see exploding stars that shine with the light of a billion suns. Just because we can't yet compute exactly how a thermonuclear flame destroys a star doesn't mean we can't measure the behavior of supernovae well enough to make them into yardsticks for measuring the size of the universe. Astronomers are used to building a case from fragmentary evidence, circumstantial evidence, and hearsay. Often there's no way to perform a controlled experiment on Earth to test astronomical theories, but we can assemble enough lines of evidence from observations to see if we're on the right path.

Most astronomy applies known laws of physics to astronomical settings, but some astronomical measurements reveal fundamental properties of the world: the underlying rules of behavior for matter and energy. Astronomical objects create settings we cannot reproduce in terrestrial laboratories.

One fundamental physical property of the world that was discovered by astronomical observation is the finite speed of light. In 1676, the Dane Ole Rømer was working in Paris, observing the moons of Jupiter. The eclipses of those moons as they ducked behind Jupiter could be predicted, but the measurements had pesky seasonal errors. Rømer had a good clock on the steady floor of the Observatoire de Paris. He noticed that in the months when the Earth's orbit around the sun brought us closer to Jupiter, the eclipses were a little early, and at other times of the year when the Earth was farther from Jupiter, the eclipses were late. Rømer inferred that light takes time to cross the diameter of the Earth's orbit. He measured this time delay to be about 16 minutes. In Rømer's time this fundamental measurement of a profoundly important physical effect—the finite speed of light—could only be done by astronomical observations. Light travels a foot in a nanosecond, a billionth of a second.[3] In the age of pendulum clocks, there was no laboratory apparatus capable of measuring such short time intervals over indoor distances. The speed of light wasn't measured on Earth until 1850, when Fizeau set up an ingenious optical device with a rapidly spinning mirror in the very same observatory. More recently, the energy and pressure associated with empty space itself is not (at least in the year 2002) detected by any laboratory experiment and is not the natural outcome of any well-established physical theory. This fundamental property shows itself only in astronomical measurements of distant supernovae that reveal an accelerating universe, which is part of the reason why this work has been so exciting.

The sluggishness of light gives astronomy, like geology, the historical reach to examine the past. We never see things as they are. We always see things the way they were when light left them. For objects in a room, that was a few nanoseconds ago. Based on terrestrial experiences, we can be excused for thinking we see things as they are. But on the astronomical scale, the effects of time ticking by while light travels are very important. They allow us to overcome our own brief lives to see how the universe has changed over long stretches of cosmic time. Light travel time transforms a telescope into a no-hokum time machine.[4] Instead of seeing a frozen moment, "now," throughout space, we see a slice through time and space: we see the present nearby, and the past when we look far away. We can trace the history of the universe by direct observation of the past, limited only by the power of our instruments.

So far, we have no way to see the future, but we can use direct measurements of the past and our physical understanding of how things work to predict the future. The stars do not predict our future, but we can predict the future of the stars, based on a firm grip of events on the scale of atomic nuclei that keep stars shining. For stars, these predictions can be tested, because we see stars of various ages, and we can trace their life cycles from birth through maturity to a death that can be quiet or violent.

The finite speed of light is woven into the language of astronomy—we use the term "light-year" to mean the distance that light travels in a year.[5] The time it takes for light to reach us from a star 100 light-years away is just a century. You can walk out tonight and see stars whose light was emitted before your parents were born. Light from the most distant supernova so far observed carries information about the way the universe has been expanding over the past 10 billion years, two-thirds of the way back to the origin of time at the Big Bang. Measuring the light from these very distant stars is not easy—the sky is bright, the stars are dim, and there are many pitfalls for the unwary—but the rewards for assembling a coherent picture of the universe are great.

In 1917, when Einstein began to connect his newly minted gravity-as-geometry with the universe, astronomers thought the stars of the Milky Way were the entire contents of the universe. Now we know the Milky Way galaxy is not the whole universe but just a small part of it. Stars form in colossal galaxies and the galaxies, each one 100 billion stars like the sun, are the units we can see that trace the underlying properties of the universe.

The sun is located in one of the outer spiral arms of the Milky Way, about 20,000 light-years from the center. All the stars you can see at night are in the Milky Way galaxy, and many of them are in that faint flattened band of light that city dwellers never see. The generous size of the galaxy means that many momentous events have already taken place, but we just carry on in ignorance because the news hasn't yet reached us. Andrew Jackson, Old Hickory, won the Battle of New Orleans in 1815, 15 days *after* the peace treaty with the British was signed in Ghent, Belgium. It took time for the message to reach him so he soldiered on until he heard the news. The flash from a supernova exploding in the Milky Way travels at the speed of light, but there is a similar lag as information travels across a great distance: there are many supernova explosions in the Milky Way for which we haven't yet seen the light. Supernovae

THE MILKY WAY
(Detail)

Figure 1.3. **In 1917, Einstein was advised that the Milky Way *was* the Universe.** Mistaking a part for the whole is common with large entities. Copyright © 2002 *The New Yorker* Collection from cartoonbank.com. All Rights Reserved.

erupt every 100 years or so in a galaxy like ours. Since the light from a supernova might take 20,000 years to travel to us, light from hundreds of supernovae in our own galaxy is on the way to us now, the flash from each one a growing shell traveling outward at the speed of light, like a ripple in a still pond from a fish leaping at twilight. Will one of those little waves lap up on our shores tonight? Will we get to see a supernova in our own galaxy, the way Tycho Brahe, the world's last great observer before the invention of the telescope, did in 1572? We don't know. We *can't* know, since no information travels faster than light to give advance warning. The last really bright supernova was seen in 1987—not in our galaxy, but in our southern neighbor, the Large Magellanic Cloud. Personally, I am ready for another one.

Individual stars are very small compared to the distances between stars, but galaxies are not so tiny compared to their separations. If you imagine a scale model where a star like the sun has the size of a pea, neighboring stars would be 100 miles away. Since

Figure 1.4. **The spiral galaxy pair NGC 2207 and IC 2163**. Distances between galaxies are not always large compared to the sizes of galaxies. These two are colliding. Note the absorption of light from one galaxy by dust lanes in the other. Courtesy of NASA and the Hubble Heritage Team (STScI/AURA).

stars are so small compared to the distances between them, they rarely collide and our galaxy seems a spacious place with a dark sky. But the distances between galaxies, although a million times bigger than the distances between stars, are not so big when compared to the galaxies themselves. If you imagine our galaxy as a dinner plate, then our nearest big neighbor galaxy, the Andromeda galaxy (also known as M31, from its place in the Messier catalog of fuzzy objects), would be just ten feet away, at the other end of the Thanksgiving tablecloth down by Uncle Bill. As galaxies move under their mutual gravitational pull, it is not rare for them to collide and possibly merge. But galaxies undergo a strange sort of collision, quite different from two plates smashing together near the gravy boat, because the individual stars that make up each galaxy are still quite unlikely to hit one another. In about 5 billion years, the Milky Way where we live and M31, now a little over 2 million light-years away but heading our way, will collide. The individual stars will miss one another, like intersecting swarms of bees.

Galaxies are distributed throughout the observable universe, with typical separations of a few million light-years. They are quite gregarious, forming loose groups and dense clusters where the galaxies crowd together, leaving large voids a few hundred million light-years across where galaxies are rare. The Milky Way is in a

Figure 1.5. **The nearby spiral M31**. M31 is part of the Local Group of galaxies. In the 1920s, Hubble observed individual cepheid variable stars in this spiral galaxy that showed it was too distant to be part of the Milky Way and must be a distant system as about as big as the Milky Way. Courtesy of P. Challis, Harvard-Smithsonian Center for Astrophysics from the Digital Sky Survey.

small group we call the Local Group that includes the Large and Small Magellanic Clouds, M31, and M33 (another nearby spiral galaxy), among others. The nearest moderate-sized cluster of galaxies is in the direction of the constellation Virgo and dubbed the Virgo Cluster. Judging distance from the apparent brightness of stars in those galaxies as seen with the Hubble Space Telescope, Virgo Cluster galaxies are located about 50 million light-years away. With a small telescope at a site with a dark sky, it's no problem at all to see these and still more distant galaxies whose light was emitted when dinosaurs still roamed the Earth.

The limit of present-day observation is the image of the "Hubble Deep Field," produced by adding up 342 images taken over 10 days at the end of 1995 with the Hubble Space Telescope. These hours of staring at a very small blank spot in the northern sky have produced

Figure 1.6. **The Hubble Deep Field**. Composed from 342 images taken over 10 days at the end of 1995, the Hubble Deep Field represents the limit of present methods for observing faint, distant, and young objects. Almost every dot and smudge in this picture is a galaxy, with light from the most distant ones traveling 12 billion light years to reach us. Courtesy of R. Williams/NASA/STScI/AURA.

our deepest image of the past. HST is in orbit above the Earth's atmosphere, so it can make images that are not blurred by the ever-changing air. But it is a relatively small telescope, only 1/16 the area of the biggest ground-based instruments, so the Space Telescope takes a long time to gather light from faint and distant galaxies. Almost everything in the Hubble Deep Field image is a galaxy. Galaxies in the foreground overlap with galaxies in the background until the Hubble Deep Field begins to show wall-to-wall galaxies. The Hubble Deep Field is the ultimate in imaging with today's technology, taking us back to the deepest accessible strata of cosmic history, within about 2 billion years of the Big Bang.

I still can call up the sharp pang of disappointment I felt at age 12 when I was working my way through the big fat volume of *The Complete Sherlock Holmes*. When Holmes walked down the path at the Reichenbach Falls for his deadly encounter with Moriarty, I felt a boyish sadness at the demise of the best and wisest man Dr. Watson (and I) had ever known. But worse was the feeling, "Is that all there is?"

And in a funny way, the Hubble Deep Field evokes a little of the same feeling. Is that *it*? Is that as far as we can see? Since we have plausible reasons to think the universe is about 14 billion years old, then the most distant thing we could possibly see emitted its light 14 billion years ago. In other words, the finite time since the Big Bang and the finite speed of light place a natural limit to our direct knowledge of the universe—the patch we could possibly observe is only 14 billion light-years in radius. Photons from some objects in the Hubble Deep Field were emitted about 12 billion years ago. So, is that *it*? Have we reached the edge of knowledge (or at least 12/14 of the edge of knowledge)?

In the same way, it is a little deflating to live in such a small and cramped universe. If the typical distance between galaxies is a few million light-years, then if each galaxy were the size of a dinner plate on a holiday table, we would reside in an observable universe only 20 miles in each direction. The observable universe seems more like crowded, jostling Hong Kong than the big sky country of Montana.

Yet *The Complete Sherlock Holmes* had another three-inch thickness of pages I had not read. This should have been a hint that Conan Doyle would relent and that there was much more Sherlock to enjoy. In the same way, a moment's thought shows there is much we have not yet read in the cosmic text. The Hubble Deep Field image was observed in colors of light that span just a slightly broader range than our eyes can see. But as we look deeper to see more distant galaxies and supernovae, still earlier in cosmic time, the light emitted from the first generation of objects in the universe would have been stretched by cosmic expansion right out of the Space Telescope's view and out into infrared wavelengths.

It's as if we have come in late to a movie. I hate that feeling. We've missed the coded messages of the opening titles and all the important early action—in the universe that's the origin of the expansion, the freezing out of helium, then the formation of the very first objects, the explosions of the very first stars, and the beginning of chemical change that makes the rich and varied world we live in, including the carbon, oxygen, calcium, and iron of our bodies. Much of this action took place even farther in the past than we can hope to see with instruments that operate at the visible wavelengths where our eyes work, Earth's atmosphere is transparent, or where HST has done most of its work.

HST is not looking in the right way to see the very first light from objects in the early universe. If we want to see the opening sequence, we will need to build an equivalent of the HST that works at longer wavelengths, in the infrared: the next-generation space telescope. And we are.

If we want to see the glow of the Big Bang itself, we need to look at even longer wavelengths of light, out where radios work but none of our senses do. And, since 1965, we have been doing that, too. But most of the universe is invisible, even with all our technical means. We know it is there because we see its effects, but we cannot measure it directly. The universe we see is controlled by the universe we do not see: dark matter that is not like the neutrons and protons that make up our bodies, and an enigmatic dark energy that shows itself in the runaway expansion of the universe.

We can build a coherent picture of the universe through astronomical observation and physical theory. Both are hard work, with many false steps, long periods of drudgery, and brief flashes of excitement. Science is not a vast encyclopedia, it is a thin flame of reason burning across ample reservoirs of ignorance. Discovering how the world works is an adventure. We may be brief and we may be short, but we are lucky enough to be here at a moment when technical advances bring new light to old human questions about the past and future of the universe. Supernovae form our method of inquiry, the dark energy is our quarry. The game's afoot!

2

violent agents of cosmic change

Peter Challis is a big bear of a guy. Sitting in the air-conditioned computer room at the Cerro Tololo Inter-American Observatory headquarters in La Serena, Chile, Pete is wearing his "Center for Astrophysics" T-shirt for the third day in a row, cargo shorts, and sneakers. He looks like he just stepped off an Ann Arbor softball diamond. It's evening and the lights of the coastal city scintillate down below. Pete isn't looking. His attention is riveted to his computer screen. Pete is making judgment calls on what he sees there.

"Junk."

"Noise."

"Binary."

Pete is sifting through images of distant galaxies, searching for supernovae as carefully as a prospector looks for the flash of gold in his pan. Brian Schmidt's fancy software has picked out candidates, but not all of them are real stars. Not even most. More like 1 in 10. Somebody has to sift the gravel from the gold. That would be Pete. The pressure is on because the high-z team Pete is playing for tonight needs some supernovae right now. Alex Filippenko is in the air, flying from Berkeley to Hawaii to observe at the Keck telescope tomorrow night. He'll vibrate to destruction without some targets to work on. I've promised to provide supernova positions to the control center at the Space Telescope Science Institute by Tuesday, just 60 hours from now. They will proceed with our plan, but if Pete doesn't find some supernovae very soon, the world's most expensive telescope will observe fields without supernovae in them. Bruno Leibundgut has time on a monster 8-meter telescope at the European Southern Observatory, up in the north of Chile, starting in 22 hours. He won't have much fun if we don't have supernovae.

An hour later, Pete's perseverance furthers our cause.

"Bingo! We got one!"

Pete's colleagues look up briefly from their computer screens in the flat fluorescent light of the computer room.

"You buy the next pizza," Nick Suntzeff says.

One supernova is good, but they need three more by morning. The only way to find them is to grind on through the night. Last night's images are gigabytes spinning on the disks, full of false alarms and a few real nuggets.

Pete keeps looking.

The universe has been changing very slowly over time, so slowly that asking your grandmother to tell you what she remembers from her childhood doesn't help to understand the aging of stars, the accumulation of heavy elements, or cosmic expansion. Supernova explosions are the exception. These violent events play out on the human timescale of days, months, and years. But even if we don't see cosmic change any more clearly than a mayfly sees a redwood age, the whole universe is changing. On the microscopic scale, the atoms that make up the stars and gas of the universe have grown more complex over time as stars fuse light elements into heavier ones to fuel their brilliance. When stars explode as supernovae, the wreckage expels fresh products of nuclear fusion into the gas between the stars.

On the big scale, galaxies mark cosmic expansion. Pete Challis is looking for evidence of this—he is looking for supernovae halfway across the universe to see how cosmic expansion has changed since the light was emitted from those distant explosions. Supernovae work well for measuring cosmic distances, but you wouldn't want to use a measuring rod you don't understand. For a long time, Pete has been part of a team trying to learn what supernovae are and how they work. The roots of these investigations go right back to the beginning of modern astronomy.

How do we know which atoms are present in the shreds of a distant star and how do we learn about motion in the universe? This is routine stuff now, but in 1835, authorities thought these things were not knowable. The French philosopher Auguste Comte declared:

> On the subject of stars, all investigations which are not ultimately reducible to simple visual observations are . . . necessarily denied to us. While we can conceive of the possibility of determining their shapes, their sizes, and their motions, we shall never by any means be able to study their chemical composition. . . . I regard any notion concerning the true mean temperature of the various stars as forever denied to us.[1]

Scientists love to quote Comte, because precisely at the time when he was making these pronouncements, the chemistry and the temperatures of stars came into the grasp of astronomy. Comte illustrates the hazards of declaring which aspects of the physical world lie beyond understanding. The zone of the unknowable has been shrinking. In the 1800s, the shrinking realm was the nature of stars; in the 1900s, the shrinking realm was the nature of the universe at large; today, the shrinking realm concerns the first and last moments and true contents of the universe, which are emerging from pure speculation into the world of observation.

Since 1704, when Newton published his *Optiks*, physicists had been clear on how to split sunlight, using a prism to form a rainbow from white light. In 1814, Fraunhofer, an optics manufacturer in Munich, used a more elegant spectroscope than Newton's to see that the spectrum of sunlight was not a continuous rainbow of color from blue to red. There were some narrow gaps in the spectrum— missing colors in the rainbow. The places where there is *no* light hold the key to unraveling the mystery of cosmic chemistry. Like detectives, astronomers gather evidence to build a picture of past events. Spectra are the fingerprints that identify elements.

A prism or grating spreads light from a star into the colors of the rainbow. The scientist's job is to take something beautiful and turn it into a graph. We plot the amount of light at each color (or wavelength) of light. What Newton didn't see, but Fraunhofer did, are the dark lines or gaps in the spectrum. The dark lines in a stellar spectrum become sharp dips in a graph and bright lines form sharp peaks in a plotted spectrum. These unique patterns identify chemical elements. For example, if you take the element calcium—found in chalk, cheese, and bones—and heat it up as Bunsen did in his burner, it gives off light at very specific wavelengths. If you see those lines, you know you are looking at calcium atoms.

Just as in the curious incident of the dog in the nighttime, we solve the mystery of the chemistry of a distant star by paying attention to places where the spectrum does nothing.[2] Calcium in a star's atmosphere absorbs light at exactly the wavelengths where calcium atoms in a terrestrial lab emit their light. Spectroscopy lets us reach across the light-years to measure the chemical composition of distant objects.

Applying spectrum analysis to the stars, beginning in the 1850s, produced a deep change in astronomy. To capture that idea, the new journal started by the American Astronomical Society in 1899 was called *The Astrophysical Journal*—in 1899, "astrophysics" meant precisely the application of spectrum analysis to astronomy. Today "astrophysics" is just a more forbidding synonym for astronomy—if an airline seats you next to a garrulous stranger and you don't want to talk, you tell them you're an astrophysicist and that usually shuts them up. If that doesn't work, you tell them you are a physicist. That always stops the conversation. On the other hand, if you're feeling expansive and you do want to chat, you tell them you're an astronomer. "Oh really, an astronomer? I'm a Leo."

The subatomic world is grainy in a way that the world of everyday objects is not. Near the positively charged nucleus, the energy of electrons is constrained to certain discrete values. It's like an elevator—you can get on and off at the floors, but not in between. Electrons take quantum leaps between states that correspond to different floors. The spectrum of an atom is set by the energy steps

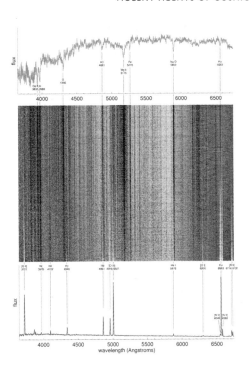

Figure 2.1. **Galaxy Spectra.** Astronomers take the light from a galaxy and spread it into a rainbow. Then they construct a graph as shown at the top and the bottom. The galaxy spectra at the top of the rainbow have absorption lines, those near the bottom have emission lines that come from gas clouds whose atoms are excited by the ultraviolet light from stars. Courtesy of Barbara Carter, Harvard-Smithsonian Center for Astrophysics.

between those grainy states—a hydrogen atom can absorb or emit only photons whose energy is the energy difference between one level and another. The observed spectrum of a star depends on the internal workings of these tiny systems.

By understanding the structure of atoms and mastering the counterintuitive rules of quantum mechanics, pioneering astrophysicists transformed the empirical world of astronomical spectra, compiled in giant catalogs, into a powerful tool for analyzing the physical universe.

This is not just qualitative knowledge, but quantitative, too. We know how much of each element is present in a typical star's atmosphere. The simplest elements, hydrogen and helium, are by far the most abundant. The next most abundant elements, carbon and oxygen, are 10,000 times rarer, and all the elements beyond helium taken together add up to only about 1 percent of the mass of a star. In the distant past, the complex atoms were even less abundant— the universe has grown richer in heavy elements over time. Second- and third-generation stars such as the sun inherited the family silver from their ancestors. Also the family carbon, calcium, and iron.

Stars are balls of gas, where outward pressure from hot gas in the interior balances the inward pull of gravity. Each star emits light at its surface, and the energy books must balance, too. If a star didn't replace the energy that it radiates away, it would shrink and wink out in just a hundred million years. In the middle of the 1800s that cooling time, 100,000,000 years, was the conventional lifetime of the sun. When Lord Kelvin, a prominent theoretical physicist, articulated this argument for the limited duration of the sun in 1862, the message was so clear and powerful that it intimidated Charles Darwin.[3]

The first edition of *Origin of Species* estimated the age of the Earth, based on geological erosion, at 300,000,000 years. Awed by the power of theoretical physics, which showed this long timescale was not consistent with the sun's lifetime, Darwin omitted his discussion of timescales from later editions and left open a serious question. Had there been enough time for his proposed natural selection to operate? Arguments from fundamental physical theory are often asserted in a loud voice with a grave tone of authority, and Lord Kelvin's pronouncements were definitely not the last occasion of this phenomenon. But what Lord Kelvin could not know was that the subatomic world discovered just at the beginning of the 1900s produces both a reliable clock for measuring the age of the Earth and a stupendous and durable source for stellar energy.

We now know the Earth is much older than Lord Kelvin declared or than Darwin estimated from the wearing down of landforms. Our clock is the very slow, but extremely steady, accumulation of radioactive decay products in rocks as one nucleus changes into another. Nuclear forces are much stronger than the electrical forces that determine the height of mountains or the bounce in baseballs. Even extraordinary variations in temperature or pressure don't affect the rate of change among the neutrons and protons of a nucleus. As a nucleus emits subatomic particles in radioactive decay, it can become another element. Radioactive uranium becomes stable lead. From the relative abundances of the parent and daughter nuclei we accumulate evidence that the Earth and the solar system are almost 5 billion years old. Darwin can relax in his grave. There has been plenty of time for natural selection to operate. From the fossil record in sedimentary rocks we know that life started simmering along at the single-cell level 3 billion years ago, and began burgeoning 600,000,000 years ago—the sun has been steadily shining for a much longer time than Lord Kelvin supposed and a good thing, too, because complex life took a long time to evolve here on Earth.

In the 1920s astronomers speculated about the origin of the sun's energy, but their estimates of stellar lifetimes were handicapped by the rudimentary state of nuclear physics. The energy source for the sun is nuclear fusion in the hot, dense core of the star. But it is a subtle chain of transformations. Deep in the sun's core, several steps of nuclear fusion transform four nuclei of the element hydrogen into a single helium nucleus. Since the sun is made mostly of hydrogen, fusion has an ample source of fuel. Unlike ordinary cooking, the mass of the assembled helium is less than the mass of the ingredients. The balance shows up as energy according to a very well-known (but not so widely understood) equation: $E = mc^2$.

More quantitatively, fusing 4.000 kilograms of hydrogen produces 3.972 kilograms of helium. Einstein's equation says you get to exchange the missing 0.028 kilogram into energy at the going rate, which is c^2. Because c is so big and c^2 immense (10^{17} joules of energy for every kilogram of mass), the energy release from nuclear fusion is astonishing. At current rates charged for electricity, pure energy has a street value of $1 billion per kilogram. Ordinary chemical reactions rearrange electrons in the outer parts of atoms, which are bound to the nucleus by electrical forces. The energy release in a candle ultimately comes from electrical forces. But nuclear reactions come from rearranging neutrons and protons in the nuclei of atoms which are 10,000 times smaller than atoms. The powerful forces acting on that tiny scale are larger: the energy released in nuclear change is typically a million times the energy released in chemical change.

Now that astronomers understand the sun's structure and composition and know how nuclear fusion yields energy, we can predict the future of the sun. We use the same authoriative tone of voice as Lord Kelvin, but this time with better understanding. The sun has ample supplies of hydrogen for another 5 billion years of steady fusion. This provides a useful upper limit to the duration of long-term financial investments.

Eventually, the accumulated ashes of hydrogen fusion, helium nuclei, begin to make a difference to the structure of a star. As you combine four hydrogen nuclei into a single helium nucleus, fewer particles barge around in the star's core to provide the gas pressure that balances gravity. A star needs to balance out the internal forces that make a star expand or shrink. About 10 billion years after it formed, that is, 5 billion years from now, the sun will adjust by swelling up to become a luminous but cool red giant star, with a diameter 100 times larger than it has today. Seen from the Earth, the sun will cover almost half the sky. The sun's florid old age will not be a pleasant era for earthlings, if there are any, 5 billion years in the future, because the Earth will heat up, what's left of the oceans will boil, first cooking all the lobsters, then melting the rocks, and eventually evaporating our favorite planet.

Our sun's elder brothers, stars similar to the sun but formed earlier in the history of our Milky Way galaxy, have already had enough time to become red giants. We see red giants of a little less than one solar mass in globular clusters, great clusters of 100,000 stars in our galaxy, with all the stars of very nearly the same age. Based on our understanding of the timescale for fusion in stars, these globular cluster stars must be about 12 billion years old. Globular cluster stars formed out of the ambient gas in our galaxy at that time. Spectra of globular cluster stars testify to the change in the chemistry of our galaxy since these stars formed. Old stars of our

Figure 2.2. **The Globular Cluster NGC 6093**. A globular cluster contains many thousands of stars that formed at the same time, early in our galaxy's history. By measuring the properties of stars that have recently become red giants (visible in the color image as reddish, bright stars in the cluster) the age of the cluster can be inferred. The oldest globular clusters have ages of 12 ± 1 billion years. Courtesy of NASA and the Hubble Heritage Team (STScI/AURA).

galaxy have only about 1/100, or in extreme cases 1/1000, the iron abundance of the sun. Something important happened between the time when the first globular cluster stars formed, about 12 billion years ago, and the time when the sun formed, about 5 billion years ago. The galaxy, anemic at first, is now rich with iron and all the other elements heavier than helium.

Red giant stars in globular clusters do not come stamped with the date of their manufacture, but practitioners in the art of determining stellar ages think the precision of this measurement for the oldest stars in our galaxy is about 1 billion years. They are willing to bet $2 to win your $1 that they are right within a billion years. That's a 1σ (one Greek sigma) result. Based on the statistics of the bell-shaped curve of probabilities worked out by the mathematician

Karl Friedrich Gauss, students of globular clusters should be willing to bet 20 to 1 that they are right within 2σ, 2 billion years. Gauss assessed the probability of getting a spurious result by chance. Rare things happen, but they don't happen very often. Gauss tells the globular cluster experts they should be willing to bet 370 to 1 that they are right within 3σ, 3 billion years. If you believe Gaussian statistics, you should be willing to bet your goldfish (4σ), your house (5σ), or your dog (6σ). In astronomy, knowing the uncertainty in a measurement can be as important as knowing the number itself because it tells you how much confidence to place in it. For important measurements, we try to give both the value and its 1σ uncertainty. But nobody really believes the statistics enough to risk their bull terrier! For the ages of the oldest stars, we write 12 ± 1 billion years, with the "± 1" intended to reflect the 1σ odds that the true answer has a different value through nobody's fault—that is, just by chance. Uncertainty is not a good thing, but knowing the uncertainty is. It keeps you from arrogance when the data are poor and gives you courage when it is warranted by the facts.

As the sun swells up to become a red giant, the energy source for the sun will shift from fusing hydrogen into helium to an elegant stage of perfect recycling where helium, the waste product of hydrogen fusion, becomes the next fuel. There is no stable nucleus with five particles. This fact of subatomic physics means there's no simple way to turn helium (which has four particles in its nucleus: two neutrons and two protons) into the next element by banging one proton into a helium nucleus. They just don't stick. So stars have trouble making the next elements, lithium, beryllium, and boron, out of helium. Instead, red giant stars skip across that gap, as improbably as crossing a stream by stepping on a salmon, to fuse three helium nuclei into a single carbon nucleus. (Carbon has 12 particles in its nucleus: 6 neutrons plus 6 protons, made from 3 helium nuclei with 2 neutrons and 2 protons apiece.) In addition, carbon and helium will fuse to make oxygen in the sun when it is a red giant.

This remarkable stage in stellar energy generation explains important astronomical phenomena through subtleties of nuclear physics. Fred Hoyle proposed it and Edwin Salpeter elaborated it in the 1950s.[4] In 1997, these two received the Crafoord Prize for this work from the hands of the King of Sweden amid trumpet blasts in Stockholm. At the dinner, the King and the prizewinners were at the center, and the guests spiraled outward in order of importance. My fiancée, Jayne Loader, was promoted toward the royal center to balance out the dearth of women among the academicians. In the outermost circle, I sat with Fred Hoyle's teenage granddaughters. I told them I was an astronomer. A Leo, actually.

After a star has made carbon and oxygen by this prizewinning process, there is still more nuclear energy to squeeze out of fusion, all the way up to iron with 56 nuclear particles. But the sun will not burn its carbon and oxygen. Only more massive stars, typically 10 times the mass of the sun, can do a thorough job of extracting all the energy from nuclear fusion.

Iron is the most tightly bound nucleus. Stars extract energy from nuclear fusion by building up heavy nuclei from light ones all the way up to iron. This makes iron the end of the road for fusion, but iron is by no means the most complicated nucleus in nature. Lead and gold and uranium are all more elaborate elements whose nuclei have more neutrons and protons than iron does. Uranium-238 has 92 protons and 146 neutrons, far beyond the total of 56 baryons for iron. Power reactors on the Earth release nuclear energy from fission—by splitting uranium nuclei into smaller pieces. In this case, the combined mass of the smaller pieces is *less* than the mass of the uranium you started with. The balance is exchanged for energy at the usual extravagant rate. So you can get energy from fusing together light nuclei up to iron and you can get energy from fission by breaking up bigger nuclei down to iron. Iron itself is the nuclear turnip out of which no more blood can be squeezed.

These details of nuclear physics affect the way stars generate energy, and they also affect the chemistry of our galaxy and of every galaxy. Lithium, beryllium, and boron are rare elements throughout the universe. They are formed by heavier elements that break up when they are whizzing through interstellar space as cosmic rays. These rare light elements are the ones skipped over by stars when they fuse helium into carbon. Carbon and oxygen are a million times more abundant. Everybody has seen carbon in graphite or coal or diamond. And carbon is the basis for the chemistry of life—at least here on Earth. Diamonds may be a girl's best friend but your best girlfriend is carbon.

Stars make the elements in accord with microscopic rules set by nuclear physics. Carbon-based life-forms like us are made of stardust whose composition is determined by subtle details of furious nuclear collisions in the centers of stars. Sometimes people look to the stars for our origins—in this very literal sense, we did come from *out there*. But not in shiny saucers. We arrived atom by atom in the gas and dust that formed the solar system 5 billion years ago. The carbon nuclei incorporated into the base pairs of your own DNA were synthesized in the fiery hearths of red giants before the sun formed.

Like loyal alumni, successive generations of stars have donated their atoms to the chemical endowment of our galaxy. While globular cluster stars had to make do with the thin gruel of the early galaxy when they formed, the sun, formed about 7 billion years later, inherited heavy elements from stars that vanished long ago.

After a brief but glorious 1 billion years as a red giant, the sun will begin to puff off its outer atmosphere, while its core hunkers down under the relentless force of gravity to become a dense white dwarf star, about the size of the Earth. During the transition, the star and its departing gas form a beautiful "planetary nebula"—an object that resembled a planet when seen in early telescopes. White dwarf stars have a small surface that doesn't emit much light, and we can see them only when they are quite nearby, as in the case of Sirius B (the white dwarf flea that accompanies the brightest star we see from Earth, Sirius, the dog star). A white dwarf is held up by quantum forces between electrons, not by gas pressure. This "degeneracy pressure" can support a white dwarf, even as it cools into invisibility. But the quantum mechanical support for a white dwarf is overwhelmed by gravity at a sharp upper mass limit of 1.4 solar masses. This upper mass for a white dwarf was worked out by Subrahmanyan Chandrasekhar (a suitably astronomical name: Chandra means "moon" in Sanskrit), and is known as the Chandrasekhar limit.

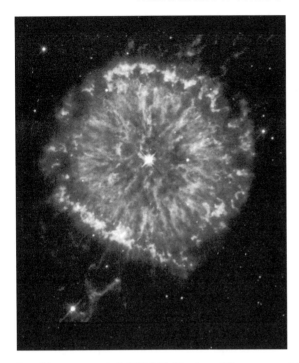

Figure 2.3. **Planetary nebula NGC 6751**. After about a billion years as a red giant, a star like the sun will puff off its outer envelope while the core shrinks to become a white dwarf. A planetary nebula is the beautiful transition from a gaseous star with nuclear fusion to a solid star with no energy source. Courtesy, NASA and the Hubble Heritage Team (STScI/AURA).

A single white dwarf emits a little light from its surface, but no longer has a nuclear furnace to replace the heat it radiates away. A white dwarf cools and fades away like a memory, slowly slipping below the edge of detection. This is the way the sun will end: it will go not with a bang, but with a whimper. Simple physical principles of heat conduction show how these faint stellar clinkers dim as they age. The coolest, dimmest, most boring white dwarfs provide a cosmic clock to compare with the globular cluster ages. The dullest white dwarfs took almost 10 billion years to cool—they appear to be just a shade younger than the oldest globular cluster stars.

This is a good result. Globular cluster experts can keep their dollars, goldfish, houses, and dogs. When astronomical measures agree, in this case, the age of the galaxy from white dwarf cooling and the ages of red giants in globular clusters, it makes you think we may be blundering toward the truth. While both arguments are complex and have uncertainties that are hard to evaluate, there are many ways to disagree but only one way to agree. It doesn't prove that both are right, but when independent paths lead to the same conclusion, there's hope that we're not just fooling ourselves.

White dwarfs with binary partners can do more interesting things than go gently into that good night. Sirius and Sirius B are locked into a dance by their mutual gravity. Closer stellar pairs can interact: white dwarfs in binaries can explode as type Ia supernovae as the white dwarf nears the Chandrasekhar limit. This won't happen to the sun, because it is an only star, but most stars are born into multiple systems where it's possible for a too-generous sibling to dump gas onto an orbiting white dwarf, precipitating a disaster. The reasons for thinking type Ia supernovae come from exploding white dwarfs are powerful, but mostly theoretical. Nobody has yet identified such a system before the explosion, and then seen it erupt. So far, observations don't show any sign of the not-so-innocent bystander being smashed by the explosion. Another possibility that is not ruled out is that type Ia supernovae come from binary systems where both stars are white dwarfs that radiate their orbital energy in gravitational waves and spiral together to create an explosion.[5] Despite our uncertainty about exactly how a white dwarf meets a violent end, we do know that stars explode, and the circumstantial evidence favors white dwarfs as the origin for type Ia supernovae.

Type Ia supernovae are found in galaxies of all types, the spiral and irregular galaxies where massive stars are forming today, and the elliptical galaxies where, as in globular clusters, most of the star formation took place 12 billion years ago. SN Ia are the only type seen in elliptical galaxies today, where the current rate of star formation is very low. This suggests that the path to becoming this type of supernova must be long and slow, as it might well be for a one solar mass white dwarf in a binary. It could easily take several billion years for a star of modest mass to use its fuel, spend some time as a red giant, and settle in as a white dwarf. If the companion is also a low-mass star, there could be a long delay before it begins to gently rain down the extra mass that nudges a white dwarf to thermonuclear destruction.

SN Ia are thermonuclear explosions—nuclear bombs with the mass of a star. When the carbon and oxygen in the interior of a white dwarf start to fuse, the reaction releases heat that speeds more fusion, powering an intense nuclear burning flame that rips through the dense little star. The flame burns much of the star all the way up to iron with such a huge release of energy that for a few weeks a single little star becomes as bright as four billion suns. That's the event we see as a type Ia supernova.

These explosions are spectacular and distinctive. Even though hydrogen is the most abundant element in the universe, spectra of type Ia supernovae don't show any hydrogen. This is a good hint that supernovae come from stars that have undergone significant changes. Type Ia supernovae get bright and then dim in a very distinctive way, taking about 20 days to reach maximum light, then declining by a factor of two in the next two weeks, and then slowly declining by about 1 percent per day for the next year and a half. Computations show that this light curve is powered by the decay of radioactive elements near iron in the periodic table that are produced when an explosion incinerates a white dwarf. More precisely, the nuclear burning in the violent destruction of a white dwarf makes radioactive nickel. This decays, with half the remaining nickel decaying every 6.1 days (the "half-life") to cobalt, and cobalt decays with a half-life of 77.1 days to stable iron. Type Ia light curves provide a nuclear-powered clock.

This is not just an idea. If SN Ia are powered by the decay of nickel to cobalt to iron, we should see the abundance of those elements change. Spectrum lines of cobalt should grow weak as those nuclei change into iron. In 1994, a Harvard undergraduate doing his senior thesis with me, Marc Kuchner, along with postdocs Phil Pinto and Bruno Leibundgut, used spectra of SN Ia to look for these

Figure 2.4. **Supernova 1994D.** This type Ia supernova (bright spot at lower left) is in a galaxy at a distance of about 50 million light years in the Virgo cluster of galaxies. For a month, the light from a single exploding white dwarf is as bright as 4 billion stars like the sun. Courtesy of P. Challis, Center for Astrophysics/STScI/NASA.

changes. We measured spectra taken in the weeks after maximum light and we found a decrease in cobalt while the abundance of iron was rising. Just as predicted. Over a time of months, we could see before our eyes the gradual transformation of one chemical element into another by radioactive decay.[6]

Type Ia supernovae are responsible for making the iron in the Earth's core, in the Eiffel Tower, and in your own blood. In the explosion of a type Ia, the star is totally destroyed. We expect these supernovae to leave nothing behind but a hot, glowing, iron-rich cloud of shredded star, emitting X-rays. Best of all for measuring the universe, the explosions are all more or less similar, possibly because they erupt in stars pressed up against the Chandrasekhar upper mass limit for white dwarfs.

If exploding white dwarfs all emitted exactly the same amount of light, then judging the distance to a SN Ia from its brightness would be a precise way to measure distances in the universe. In fact, there is a range of energy emitted by SN Ia explosions, and we have been working hard to understand this variety. Over the past decade, these efforts to improve the precision of SN Ia as cosmic rulers have paid off: supernovae are now the best tools for measuring distances to other galaxies. These are the objects Pete Challis was so desperately seeking in the La Serena data room.

In 1983, I was on the astronomy faculty at the University of Michigan. Very early one October morning, I was awakened by an excited predawn telephone call from the *Ann Arbor News*.

"Have you heard about the Nobel Prize?"

It didn't seem possible. What had I done to deserve this? I honestly couldn't think of anything. This was terrible. Maybe I had done something wonderful, but now I had early-onset Alzheimer's and I couldn't remember what it was. Why hadn't they called me sooner, when I could appreciate it? I sat up in bed, sweating uncontrollably. Luckily, I was too groggy to say anything, and the reporter's voice pulled me out of this inward spiral of self-delusion.

"They gave the Physics Prize to Willy Fowler and, how do you say this name? Chan-dah something something," the reporter went on. "What do you think of that?"

"Oh!. . . Oh, that's great. CHUN-druh-shay-khur. Aspirate the hard k and all the a's are schwas." I stalled for time, slowly regaining brain function. It was OK. I hadn't done anything, but at least I knew it.

"It's like moonlight," I said.

"Hunh?" the reporter interrogated deftly.

"We all bask in the reflection! You see these are two of the guys who figured out how to apply nuclear physics and quantum mechanics to stars. The Chandrasekhar limit for white dwarfs, for example, ..."

This is a complex story: White dwarfs, binaries, runaway fusion of carbon and oxygen in a degenerate star, radioactive power from nickel to cobalt to iron decay, and the total destruction of a star. Is there a way to test whether this picture is correct? There's no hope of a laboratory test for the whole complex set of events, but if this story is right, we should see the essential ingredients by observing supernovae.

Most supernovae we see are in very distant galaxies: the information we can gather is limited by the object's faintness. If you look at thousands of galaxies, you can finds dozens of supernovae each year. Rare things do happen. If our galaxy is like other galaxies, events we see in distant stellar systems have corresponding, if infrequent, events nearby. If you limit your attention to the Milky Way, you must wait for centuries, but an exploding star in our own galaxy, just a few thousand light years away can be an astonishing sight.

In 1572, before the invention of the telescope, the not-yet famous 24-year-old Danish astronomer Tycho Brahe, reported the most recently observed SN Ia in our galaxy.

> On [11, November 1572] a little before dinner . . . and during my walk contemplating the sky here and there . . . in order to continue observations after dinner, behold, directly overhead a certain strange star was suddenly seen, flashing its light with a radiant gleam and it struck my eyes. Amazed, and as if astonished and stupefied, I stood still, gazing for a certain length of time with my eyes fixed intently on it and noticing that same star placed close to the stars which antiquity attributed to Cassiopeia. When I had satisfied myself that no star of that kind had ever shone forth before, I was led into such perplexity by the unbelievability of the thing that I began to doubt the faith of my own eyes, and so, turning to the servants who were accompanying me, I asked them whether they too could see a certain extremely bright star when I pointed out the place directly overhead. They immediately replied with one voice that they saw it completely and that it was extremely bright. But despite their affirmation, still being doubtful on account of the novelty of the thing, I enquired of some country people who by chance were traveling past in carriages whether they could see a certain star in the height. Indeed these people shouted out that they saw that huge star, which had never been noticed so high up. And at length, having confirmed that my vision was not deceiving me, but in fact that an unusual star existed there, beyond all type, and

marveling that the sky had brought forth a certain new phenomenon to be compared with other stars, immediately I got ready my instrument. I began to measure its situation and distance from the neighboring stars of Cassiopeia, and to note extremely diligently those things which were visible to the eye concerning its apparent size, form, color, and other aspects.[7]

As day faded into night, and day came again, Pete Challis's list of supernova candidates grew. There were no country people traveling past in carriages to check his work. Carefully screening the images on his monitor, Pete was finding something more valuable than flecks of gold. Pete Challis was picking out supernovae from images of distant galaxies, taking a step toward understanding the history of cosmic expansion. When dawn comes and he hands over his list, other members of the team will jump into action. They will gather the light from Pete's distant discoveries, spread each one into a spectrum, and note extremely diligently things that are invisible to the unaided eye. The spectrum will reveal each supernova's contribution to the stock of heavy elements in a distant galaxy and form the basis for a scientific prophesy of future cosmic expansion.

3

another way to explode

Curiously, nature has contrived more than one way to destroy a star. Both types of stellar explosion emit comparable amounts of light, so supernova types have been confused from the early days of this subject. SN Ia come from exploding white dwarfs. But other stars explode by collapsing. The idea that supernovae have their origin in collapsing stars was proposed by Fritz Zwicky and Walter Baade in 1934. As explained by Willy Fowler and Fred Hoyle in 1960, stars with eight or more times the mass of the sun do not produce white dwarfs at the end of stellar burning, but have a different way to explode. For massive stars the explosion energy comes from gravity, not from fusion. Although massive stars have different histories, different structures, and a different energy source for the explosion, the light that is emitted is not so different, so it has taken decades to sift out gravity-powered supernovae from their thermonuclear cousins. This is very important if you want to estimate the distance of a star by using its brightness. To get good results you had better compare objects that are the same. If you don't recognize all the various types of supernovae, you will be sure to make errors in the distances.

Massive stars burn their fuel more quickly than low-mass stars. A star with 10 times the mass of the sun has ten times the fuel to burn, but uses its fuel 10,000 times faster to shine 10,000 times more brightly than the sun, so it exhausts its nuclear energy supply one thousand times faster. Quantities matter: 10 million years for the duration of a 10 solar mass star is very different from 10 billion years for the sun's lifetime in the same proportion as a ten-dollar bill is different from a penny. Ten million years is short. For a star.

Though they are brief, massive stars are thorough. Massive stars squeeze energy from nuclear fusion from carbon and oxygen into silicon and sulfur and then all the way up to iron. Most of the star is still unburned hydrogen, but the interesting stuff is hidden deep within the star's core, where helium and the heavier elements reside. The residue from hydrogen burning is helium, the ashes from burning helium are carbon and oxygen, oxygen burning produces elements near silicon, and the fusion of silicon reaches the dead end of fusion: iron. The products of each stage of nuclear fusion surround the iron core like the rings of a tree stump as the core relentlessly continues on its path toward destruction.

At the point where it has accumulated an iron core, a massive star is like a teenager with a credit card. It has a huge outflow, but no source to maintain its balance—for a star, that's the pressure balance against gravity's relentless inward pull. In low-mass stars, quantum mechanics intervenes to keep 1.4 solar masses of cold carbon and oxygen from collapsing, but massive stars employ pressure from hot gas to balance gravitation. As the core shrinks, trading gravitation for heat in the way Lord Kelvin imagined, the core's temperature rises.

In previous burning stages, as when a massive star ignites its carbon, a higher temperature ignites a new fuel whose energy release maintains a new, if limited, period of equilibrium. When the core is iron, this pattern ends, because you don't get any energy from making heavier elements out of iron. The star has tremendous energy flowing out of the core, much of it in the form of deadbeat neutrinos that have no electric charge and don't bounce off nuclei either, so they stream out freely and contribute nothing to the support of the overlying material. Eventually, the central temperature reaches 3 billion kelvins at which point the iron nuclei begin to melt back into lighter nuclei.[1] This doesn't produce new energy—it costs energy to break up iron. The inevitable then takes place. The core, about 2 solar masses with a radius about half the size of the Earth, loses its pressure support and suddenly slumps inward. Gravity is so strong in the dense small core that this implosion takes only one second as the iron core accelerates inward to about a third of the speed of light. As the inrushing core approaches the density of an atomic nucleus, the strong nuclear force suddenly halts the contraction and the innermost core begins to form a neutron star. This abrupt deceleration, like a train hitting a wall, sends a powerful shock wave back upstream through the imploding star and, with aid of a blast of neutrinos, ejects the outer layers of the star in a type II supernova (SN II).

Neutrinos are produced copiously just outside the nascent neutron star, about 100 kilometers from the center of the collapse. In models for SN II, the explosions of massive stars, most of the energy of the collapse comes out as neutrinos, about 1% goes into the motion of the exploding star, and only about 1/10,000 of the energy goes into the display of light that makes us pay attention to an exploding star. Although they have no electric charge, and nearly no mass, neutrinos carry energy, and this hail of energetic neutrinos plays a decisive role in making the rest of the star explode. Computer models of exploding stars (often done at weapons labs like Los Alamos or Livermore, which have a professional interest in physical situations where the sudden release of energy blows things apart) show that the hot gas outside a forming neutron star could well be one place where new elements are synthesized right up to the end of the periodic table. Even though it *costs* energy to make iron into gold, the region just outside the nascent neutron star is made of iron and there is lots of energy from the powerful shock wave ripping through the star. Supernovae turn iron into gold, gold into lead (oops!), and lead into uranium. Elements beyond iron are rare in nature because they are made in very special environments.

Massive stars also blast off their thick unburned and partially burned outer layers as part of the supernova explosion. So core-collapse supernovae from massive stars will eject hydrogen if the star still has its outermost layers, and large amounts of oxygen and other middleweight elements in any case. Massive stars that exploded more than 5 billion years ago are the source of oxygen atoms that we're breathing right now.

In the 1930s, Fritz Zwicky and Walter Baade started the modern study of supernovae. They worked as a team, with Zwicky at Caltech in Pasadena, California, and Baade just a mile up Lake Avenue at the Santa Barbara Street offices of the Carnegie Institution's Mount Wilson Observatory. Baade and Zwicky coined the name "supernovae" to distinguish them from ordinary novae. Novae are explosions on the surface of a white dwarf that are 10,000 times dimmer and do not destroy the white dwarf. Supernovae, rarely seen in our galaxy, but more frequently when you search large volumes that contain many galaxies, are much more violent. Although Baade and Zwicky discussed the nugget of this idea in a legitimate scientific setting, a meeting of the American Physical Society, the most vivid early form of publication was a cartoon in the *Los Angeles Times* on 19 January 1934.[2]

This is one of Zwicky's remarkable insights, perhaps second only to his discovery of dark matter in galaxy clusters. Zwicky's impact on astronomy has grown over time as supernovae and dark matter have bubbled to the top of the astronomical stew. Fritz died in 1974, and a Ph.D. takes about five years, so five generations of astronomers have grown up knowing the legend but not the person. For those of us who actually met Fritz, as I did in early-morning encounters in the second sub-basement of Caltech's Robinson Lab, time has begun to erode and soften the memory of his abrasive personality. Somewhat. What remains are the ideas without the person: in Zwicky's case this has made it easier to admire his work.

Looking back, we see Zwicky and Baade bravely attributing the energy in supernova explosions to a wild idea: the gravitational collapse to neutron stars. And, in the years after this insight, Zwicky built the first telescope at Palomar Mountain, the 18-inch Schmidt, to follow up this idea, backing his talk with action. Since we now know that some supernovae are, in fact, powered by gravitation and do indeed leave neutron stars, we credit Fritz with another daring insight.

Truth is more complex than legend. Zwicky, working at Caltech, had begun the systematic study of supernovae to check his theory of collapse to a neutron star. Fritz discovered one supernova in 1936 and six in 1937. All of the supernovae that Zwicky and

Be Scientific with OL' DOC DABBLE.

Figure 3.1. **Be Scientific with Ol' Doc Dabble.** Zwicky's compact 1934 publication of a wild speculation for the origin of supernovae in the gravitational collapse of stars to form neutron stars: "little spheres 14 miles thick." This is now thought to be the mechanism for type II supernovae, though, in 1934, Zwicky was talking about type I supernovae. Courtesy of the Associated Press.

Baade studied in those years showed very similar spectra and light curves. It wasn't until 1940 that Rudolph Minkowski, also working at Mount Wilson, observed the spectrum of a supernova that was completely different. At that point, supernovae were, quite sensibly, split into two types: type I, the original type, and type II, the new kind.[3] The legendary insight that supernovae make neutron stars was the inspiration for Zwicky's own observational work on supernovae in 1936 and 1937. But, as luck would have it, *all* of those were supernovae of type Ia—the type that does *not* form neutron stars. Sometimes a good story is better than the facts. Or, as the newspaperman says in *The Man Who Shot Liberty Valance*, "When the legend becomes fact, print the legend."

The type Ia story with degenerate white dwarfs and crenelated nuclear burning flames is complicated, but the mechanism for type II with a core collapse, bounce, and emerging shock wave seems downright baroque. How can we test whether massive stars really do all the things that the computers at Los Alamos and Livermore predict? There's no way to do a controlled test of a supernova out in the desert near Las Vegas. Astronomy is an observational science, which means we need patience, good luck, and many lines of evidence to test our ideas.

Massive stars that become SN II mature and explode so quickly they could erupt right in the cloud of gas and dust where they formed. The galaxy nearest to our own, the Large Magellanic Cloud (LMC), has many patches of lively star formation, including the giant 30 Doradus region where hot young stars make the surrounding gas glow by ripping off their electrons. The LMC is part of our galaxy's entourage—it is a satellite of the Milky Way but only easily visible from Earth's southern hemisphere. The brightest stars in the LMC have the luminosity we expect from stars around 20 times the mass of the sun. Back in the 1960s, Nick Sanduleak of Case Western Reserve University compiled a catalog of the bright stars in the LMC. One of them is not there anymore.

That star, Sanduleak −69 202, was last seen shining brightly in late 1986—as a massive blue supergiant in the LMC. But that star exploded 165,000 years ago, and emissions from the supernova explosion reached the Earth at 7:36 Universal Time on Monday, 23 February 1987. That was supernova 1987A.[4] Neutrinos, nearly massless particles with no electric charge, erupted from the forming neutron star in SN 1987A, arrived and flashed through the Earth, which is transparent to neutrinos, before anyone saw the star start to brighten.

At the Carnegie Institution's Las Campanas Observatory in the north of Chile, around 2 A.M. (5 hours Universal Time on Tuesday, 24 February), telescope operator Oscar Duhalde took a break at the 40-inch Swope telescope, leaving the astronomers in the data room, and going downstairs to heat water for his nightly coffee. While the kettle warmed on the hot plate, he stepped out for a glance at the sky. It was a wonderful clear night, the kind when astronomers can measure the brightness of stars without fear of clouds confusing the data, the kind of night astronomers call "photometric." Looking to the South, Oscar saw the large fuzzy patch of the LMC. Right near 30 Doradus, a patch of star formation in the LMC, Oscar saw something new, a bright star he'd never seen before. Neither had anyone else.

He knew this was worth mentioning to the observers, Barry Madore and Robert Jedrzejewski, but when he came into the control room, they were just reaching the punch line of an off-color joke. By the time they explained what was so funny about Italians by converting idiomatic English into Chilean Spanish, Oscar had forgotten about the new star. Barry turned up *Echo & the Bunnymen* on the sound system and they all got back to work.

Ian Shelton, a young Canadian astronomer working at the University of Toronto's telescope, also on Las Campanas, came into the control room at 4 A.M., a little like Tycho seeking confirmation from the "country people who by chance were traveling past in carriages." Ian had discovered a big solid dot on his photographic plate of the LMC, near 30 Doradus. There wasn't any star there on his earlier plate of the same place. He went outside and saw it with his eyes, but he still wanted confirmation of this nova in the LMC.

"Oh yes," said Oscar. "I saw it. Two hours ago. Near 30 Dorado. I saw it."

"A nova?" Barry, an expert on the distance to the LMC, thought for a moment, doing the inverse square computation in his head. "No," he said, "that would be a supernova."

This event was SN 1987A, the brightest supernova seen since 1604.

Theory predicts that most of the energy of a core-collapse supernova comes streaming out as nearly massless, chargeless neutrinos. One of the most interesting observations of SN 1987A was not made with a telescope, but with an underground tank of water designed to find out if protons are immortal, or just very long-lived. The experimenters had hoped to detect flashes of light produced by the death of protons inside the tank and measure a finite lifetime for the proton. This would have been quite interesting, since the physical world we see around us is made of protons. It would have shown that matter is evanescent—just a phase that nature is going through. The decay of protons was predicted by interesting theories of particle physics called grand unified theories that unite the strong and weak and electromagnetic forces in a single conceptual framework. The theorists were so persuasive that experimenters excavated a chamber in a salt mine and built a giant tank containing 6000 tons of ultrapure water to confirm those predictions. They didn't. Just before the Department of Energy cut off their funding, a blast of neutrinos emitted from the star's collapsing core flashed through their detector. This was the sharp yelp of a neutron star being born deep in the heart of SN 1987A.

When the supernova was discovered, by its optical emission, the report came to Brian Marsden, the person behind the Central Bureau for Astronomical Telegrams. His office is about 200 feet from mine, but I did not hear about SN 1987A from him. Craig Wheeler called me from Texas. A Texas graduate student was in Toronto, where everybody was talking about Ian Shelton's discovery. The student called Craig, and Craig called me.

"Bob, there's a supernova in the Large Magellanic Cloud."

"Ha, ha, ha, Craig Wheeler! Fool me once, shame on you; fool me twice, shame on me."

Nine years earlier, Craig masterminded a practical joke, sending a fraudulent urgent telegram to me in a remote village in Italy. "Return at once! Bright supernova in M51," the fake message said. I was in the midst of complicated airline ticket changes when Craig and his co-conspiritors took pity and let me in on the joke. Had I forgotten? No!

"No, no, no. This one is for real!"

Craig started to fill me in on the details. I cut him off.

"Craig, tell me all this later. Maybe we can observe this puppy with IUE. Hang up and I'll see if we can get NASA going on this."

IUE was the International Ultraviolet Explorer, a nimble little satellite that could observe at ultraviolet wavelengths where the Earth's atmosphere is opaque. I had sent in a "Target of Opportunity" proposal to observe any bright supernova that came along. Since this was the brightest in 383 years, I was pretty sure they would approve the request and aim the satellite at SN 1987A. But I didn't want to waste any time. If we acted fast, we might see ultraviolet light from the hot expanding surface of the star right after the powerful shock wave from the star's core blasts through.

I was looking up the telephone number for the Goddard Space Flight Center when my phone rang. It was Yoji Kondo, the IUE Project Scientist at Goddard. Yoji was courteous, but wildly excited. His mood was catching.

"Bob, good morning."

"Good morning, Yoji." I bowed slightly toward the receiver.

"Perhaps you have heard about the supernova in the Large Magellanic Cloud."

"Yes, I have just been speaking with Craig Wheeler who informed me of this event."

"We thought you might be interested in making observations," Yoji said.

"Yes, I think that would be of interest. "

"They have already begun."

"Yahoo!"

My partner in this work, George Sonneborn, a NASA scientist at Goddard Space Flight Center, was at the IUE control console to make these prompt observations of SN 1987A with the IUE. Our data showed the outer layers of the star being blasted off at 30,000 kilometers per second, 1/10 the speed of light. Over the next weeks, the supernova cooled and faded from our view in the ultraviolet, but we still saw *two* bright hot stars at the site of the explosion. This was puzzling. Sanduleak −69 202 was known to have one close blue neighbor. Perhaps both Sanduleak's star and its dimmer neighbor had survived and that's what we were seeing with IUE. Perhaps the star that exploded was yet another star in that crowded neighborhood of the Large Magellanic Cloud.

For a few weeks in 1987, I wasn't sure whether Sanduleak −69 202 had really been vaporized and I said so in public places. Stan Woosley, a supernova theorist at the University of California, Santa Cruz, wasn't persuaded. The match between his models and the observations was far too good. Stan said, "If it wasn't Sanduleak −69 202, the star that exploded was exactly like it." Luckily, I did not publish my mistaken conclusion, though I talked about it enough to richly deserve a roast crow, with stuffing and cranberry dressing. Careful measurement of old data showed that there had been not one but *two* additional hot blue stars there all along, hidden in the glare of Sanduleak −69 202. The IUE was seeing those other two stars. Star 202 had, in fact, disappeared. Nick Sanduleak was fond of showing a Cleveland newspaper headline drawing the conclusion, "Sanduleak Explodes!" This case of mistaken identity didn't do any permanent harm to human understanding, especially because our observational "fact" didn't convince Stan Woosley that his models were wrong. But this was an experience I did not want to repeat.[5]

On that exciting Tuesday in February 1987, I had recently moved to Harvard from the University of Michigan. At Michigan, several people in the Physics Department were part of the Irvine–Michigan–Brookhaven experiment to find the decay of protons. Since they *hadn't* found the lifetime of the proton, I thought it was my duty to call them up to alert them to a possible neutrino blast from the supernova in the LMC. I called the Michigan Physics Department. It was a strange encounter: everybody I called was in Moriond, France, at a ski resort for a very important conference on cosmology and particle physics. Undoubtedly they were studying the effects of powder snow on the gravitational descent of physicists. After 20 minutes of finding nobody home, I just left a message.

"Not the lifetime of the proton, but the supernova of a lifetime—look for the neutrinos."

Fortunately, the neutrinos had also left a message on their data-recording equipment at the mine. The team found a flash of neutrinos that had entered the tank (after passing through the Earth!) in the hours *before* the optical discovery of the supernova. A similar detector in Japan, which had been looking for neutrinos emitted by nuclear reactions in the center of the sun, saw the same event. Having two independent measurements gives you confidence that you are observing something real, not noise in the equipment.

John Bahcall was visiting Harvard from the Institute for Advanced Study in Princeton. He came to my office, looking for conversation about supernovae and for a razor. He had arrived that morning without shaving and he wanted to clean up before the physics colloquium that afternoon. John was going to bring us up to date on the puzzling measurements of neutrinos from the sun, which showed only about one-third of the amount predicted. John was doing the predicting, and he wanted to convince us that the discrepancy was real and not the result of something he might have forgotten. Like his razor.

I keep a razor in my desk, since I sometimes arrive on overnight flights from observing in Chile in an unkempt state. John used it. Clean-shaven and clear-headed, John started thinking about SN 1987A, the optical observations, and the neutrino signal. By the end of the day, after talking with his friends in the physics department, John sent a letter to *Nature*, the science journal that believes it is the world's most prestigious, using the timing of the neutrino arrival to place a stringent limit on the mass of the neutrino, better than any limit from terrestrial laboratory work in 1987. In 1999, measurements of solar neutrinos emitted from the sun's core, and from atmospheric neutrinos, both detected by giant underground water tanks, now indicate that the mass of the neutrino is not quite zero, a very important fact for particle physics, and one small source of dark matter for cosmology.

The explosion of supernova 1987A in the Large Magellanic Cloud was the best opportunity in four centuries to study the collapse of a massive star. Underground detectors in Ohio and in Japan were jolted by a sudden spike of neutrinos, signaling the birth of a neutron star in the center of the dying star. Since we know that a spinning neutron star lies at the center of the Crab Nebula, a supernova recorded by the emperor's astrologers in the Sung dynasty of China on 4 July 1054, it was natural to think that SN 1987A might have one too, so eager research groups began to look for the telltale flashes from a dense spinning nugget at the center of the cataclysm. Sure enough, in 1989 a group led by Jerry Kristian at the Carnegie Observatories and including Rich Muller, Carl Pennypacker, and Saul Perlmutter from Lawrence Berkeley Lab reported seeing the pulses at 37σ—the firm signature of the youngest neutron star ever seen.[6] If you would bet your house at 5σ, you should probably be willing to bet your life at 37σ, but nobody takes statistics that seriously. Plus, there are ways to go wrong that statistics don't include.

I was invited to give a talk in April 1989 at the National Academy of Sciences in Washington, D.C. The academicians often invite people too junior to be elected members of the Academy, but who are working on interesting new developments to amuse them at their annual meeting. It was the first time I had been to that temple of science. I was amazed by how old the academicians were. Scientific research must be good for your longevity. (Now that I have been a member of this geriatric organization for a few years, the antiquity of its members still makes me feel like a kid—maybe that is the secret of the members' vitality.) Descending the stairway to the talk, Frank Press, the President of the Academy and the father of Bill Press, one of my astronomy colleagues at Harvard, confided that he was especially interested in hearing more details about the amazing neutron star in the center of SN 1987A. According to Kristian et al.'s report published in *Nature*, the neutron star was spinning at a rate of 1968.629 times per second, compared to the Crab pulsar's leisurely 33 times a second. The investigators said they were further analyzing the data with tantalizing hints that the pulsar might be in an 8-hour orbit around an unseen companion, perhaps a planet. This was wild and exciting stuff.

However, I disappointed Frank Press. I mentioned the pulsar data, but I didn't say too much about it in my talk, because unlike the neutrinos seen in Ohio and in Japan it didn't seem to have the converging lines of independent evidence that make a scientific result secure. At the risk of seeming a dull fellow, I thought it was better to emphasize things that were interesting and true at the expense of things that were just interesting. The pulsar, though reported with great precision in *Nature*, was seen on only one night, 18 January 1989. On other nights, the same equipment and the same analysis failed to detect this amazing object. In general, if something is real, the evidence gets stronger over time. In this case there was always the possibility that the expanding clouds of debris might have allowed only a brief peek at a real phenomenon. Still, when others tried to measure the pulsar, they came up blank. This was a bad sign. If something is real, another team with an equivalent technique ought to be able measure the same thing. Having more than one group measure anything important is more than just a good idea. It makes the case.

During 1989, this mystery deepened. Was there some flaw with the original observation, even though the statistics of the initial measurement had seemed so clear-cut? Eventually, the group that had made the measurement got to the bottom of their own problem. The signal that had seemed so certainly the signature of a spinning neutron star in the center of supernova 1987A was, alas, generated in the circuitry of the television camera used to guide the telescope during the data-taking. On the night when that team made the "discovery," the TV camera was *on* while they were taking supernova data, but as dawn approached, they turned it *off* while taking calibration data to avoid damaging the sensitive TV camera. So the signal was in the supernova measurements but not in the calibration data they used to check for spurious noise. Ouch! There are many ways to go wrong. The wonderful thing about science is that eventually nature tells you when you are fooling yourself. Real objects can be measured again or measured by somebody else—false signals will eventually be weeded out.

So, is there or is there not a neutron star at the center of supernova 1987A? We still don't know. Even though the neutrino signal was just what was predicted from a forming neutron star, there isn't yet any clear evidence for one in the supernova debris.[7] One possibility is that some of the inner debris fell back on the neutron star and pushed it over the upper limit for those objects (somewhere around 3 solar masses). In that case, gravity would win decisively and the stellar core would collapse to become a black hole. A black hole is a region of space where gravitation is so strong that not even light can escape. Even so, invisible objects can have visible effects—and a black hole could have material in orbit around it that we could measure.

The site of the SN 1987A explosion can still be studied over a decade later with the Hubble Space Telescope (HST). My research team, the Supernova INtensive Study (SINS), has been observing SN 1987A since the launch of HST. The bright 20 solar mass star,

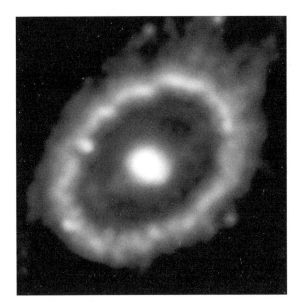

Figure 3.2. **Supernova 1987A**. Space telescope image of the site of SN 1987A, seen 10 years later. The exploded star itself is the dot in the center of the bright inner ring, heated by the decay of radioactive elements produced in the explosion. The inner ring is gas lost from the pre-supernova star, excited and still glowing from the light of the outburst. This ring was the source of the emission seen by the International Ultraviolet Explorer satellite in 1987–88. Courtesy of P. Challis and the SINS collaboration, Harvard-Smithsonian Center for Astrophysics/NASA/STScI.

Sanduleak −69 202, is definitely absent. At its site, glowing remains of the explosion are visible. It is difficult to study SN 1987A from the ground, because light from the two enduring neighbor stars (the same ones that caused me so much grief in 1987) slops into the light from the supernova. The nearby stars are 100 times as bright as the supernova is today, and from the ground, atmospheric blurring smears them into big patches of light that obscure the supernova itself. Debris from the exploded star is now 10 million times fainter than it was in 1987 when Oscar Duhalde saw it with his naked eye in Chile. It is still glowing because the explosion produced fresh elements, some in radioactive forms that continue to excite the debris. The present source of energy for SN 1987A is the decay of radioactive titanium, which will keep the debris glowing for decades. But careful examination of the SN 1987A images and spectra taken by the SINS team carried out by Harvard undergraduate Jenny Graves for her senior thesis does not reveal emission from any condensed object in the center.

Though nobody lives that long, and written records take us back only a few thousand years, we now have a good idea of how stars change on timescales of hundreds of millions of years. They synthesize heavy elements out of light ones and blast the products into the gas between the stars. This becomes the material out of which new, richer stars form. All of this stellar change takes place in galaxies, which are themselves part of an evolving universe.

4

einstein adds a constant

Stars are giant places that produce microscopic change. Through stars, the atoms of the universe have become more elaborate over time: stars knit protons and neutrons into more complex nuclei. Calcium, iron, oxygen, and carbon have increased 1000-fold in the last 12 billion years, judging their abundance in the past from spectra of the oldest stars in our galaxy. A mix of type I and type II supernovae erupting over several billion years, plus the contributions of less spectacular stars, produces the chemical abundances of the solar system. Most of the enrichment of our galaxy took place early in its history, before the sun formed 5 billion years ago from a typical scoop of this nutritious cosmic soup. The gas in the Milky Way today is only a little richer in heavy elements than the sun is because stars were forming (and exploding) more vigorously in the first 5 billion years of the Milky Way's history than in the last 5 billion years.

Spectra tell us about chemistry, but they also can reveal motion. An analogy with sound may help. We've all heard the characteristic pattern of sound as a car zooms by on a highway. Imagine a lonely hitchhiker as cars pass by on the interstate. The hitchhiker hears "ZOOOOOM," not just a shift in the loudness as the car gets closer, but a definite change in pitch from high to low just as the car zooms by. A shift in pitch from high to low tells you, even with your eyes shut, just when the car switches from approaching you (with the hope, no matter how slim, that it might stop) to receding from you. The driver doesn't notice any change as he blasts by you, just the steady hum of the engine and wheels, and perhaps a little peripheral blur of your outstretched thumb.

The shift in the apparent pitch of sound produced by a moving source is called the Doppler effect. This was proposed by Christian Doppler in 1842, in a paper at the Royal Bohemian Society for Sciences in Prague. The technology of the steam engine made it feasible to test the Doppler effect. In 1845, a skeptical Dutchman named Christoph Ballot set out to refute Doppler's theory. He placed trumpeters on a railroad car, and assembled musically trained listeners next to the track. Contrary to his expectation, Ballot's listeners heard a change in pitch, about as big as the step from one key on a piano to the next, as the trumpeters swept by.

To us, the Doppler effect is common sense, but that's only because we're used to machines that move at a noticeable fraction of the speed of sound. An 18-wheel Freightliner on Interstate 80 in Nevada is hauling along at 12 percent of the speed of sound, and its pitch drops down the equivalent of four keys on the piano as it blasts by. You no longer need to be a musician to detect the Doppler effect. The Doppler effect probably was *not* common sense for Cro-Magnon Man. Cro-Magnons didn't have highways, trucks, or trumpets.

However, we're just like our ancestors when it comes to the Doppler effect for light—that is not common sense for us. The wavelength of an atom's emission is a steadier source than any trumpeter can aspire to be. The wavelength, which we perceive as color, for an atom's emission or absorption lines is shifted a little to the blue if an atom is approaching and shifted a little to the red if the atom is moving away. But the speed of light is one million times the speed of sound, so for the same speed, the shift is a million times smaller for light than for sound and lies well below the threshold to detect a color change—even for Martha Stewart. That's why this effect is part of legend for lonesome railroad whistles, but not for their headlights. Everyday objects do not zoom by at an appreciable fraction of the speed of light, so the Doppler effect for light isn't a common sense phenomenon.

Astronomers measure the velocity of a star from the shift produced in the wavelengths of its absorption or emission lines. Here's the recipe: Gather light with a telescope, spread it out into a spectrum with a prism or grating, then carefully measure the wavelengths of the lines. Compare the measured wavelengths with the wavelengths from identical atoms, say of calcium or any other element, measured when the atoms are sitting quietly in a flame in your laboratory. The shift in wavelength tells the speed. Stars in the Milky Way galaxy have speeds measured this way of a few kilometers per second up to a few hundred kilometers per second as they mill around randomly like sailboats before the starting gun, or as they systematically orbit the center of our galaxy.

In the opening years of the 1900s, the Milky Way galaxy *was* the known universe. So if you were Albert Einstein in 1917, and you consulted your favorite astronomer about motions in the universe, the astronomer (in Einstein's case, Willem de Sitter, professor of astronomy at the University of Leiden in the Netherlands) could confidently tell you that spectra show the stars have relatively small speeds and not much pattern to their motion. This is true, but because the Milky Way is *not* the whole universe, it led Einstein down a legendary path of error and regret.

The present-day image of our galaxy as one among billions of similar galaxies was not the common-sense view or even the prevailing view among experts when Einstein was young. By counting stars in the Milky Way, astronomers hoped to gauge the extent and shape of the system in which the sun is embedded. But dust between the stars made this a treacherous undertaking. In some directions, the counts of faint stars thinned out because there really were fewer stars, so astronomers correctly inferred we lived in a flattened, disklike system. But in other directions, the star counts fell off with distance because the light of these stars was absorbed by intervening interstellar dust, distorting our true location in that disk. Dust is always a bugaboo in astronomy.

The result was a 1900s view of the Milky Way in which the sun might as well be at the center as any other place, and in which the Milky Way might possibly be the whole universe. If you're in a boat in a fog, it always looks like you're at the center of things. The

Figure 4.1. **The Milky Way**. This image shows dust clouds silhouetted against the bright bulge at the center of our galaxy. The dust makes the bulge look both dimmer and redder, as interstellar dust removes more blue light than red light. Courtesy of Axel Mellinger.

cosmic fog was absorption by dust that gave the illusion of the sun sitting centrally in an extended, flattened system shaped something like the grindstone in an old mill. The small velocities of the stars seemed to show that this whole system was neither expanding nor contracting, but just sitting there, inert and unchanging. Yet by 1930, every element of this picture was completely reversed—our location far from the center of our galaxy was clearly established, the Milky Way was seen to be just one of a huge number of equivalent galaxies, and the whole cosmic fabric was observed to be stretching out.

Just as the journey inward led to an understanding of atomic nuclei and the source of stellar energy by the 1930s, the journey outward cleared away the fog of misunderstanding about our location and the state of the universe. In 1916, Albert Einstein was trying to understand how gravity works in the universe. After his great success in 1905, creating the theory of relativity, inventing the photon, and demonstrating the reality of atoms, he was no longer a technical expert third class (with provisional appointment) working at the Swiss Patent Office, revolutionizing physics in the moments snatched between inspecting dynamo designs and rejecting perpetual motion machines. By 1915, Einstein had been transformed into Herr Professor Doctor in Berlin at the Kaiser Wilhelm Institute, where he was struggling to construct the mathematical structure of his theory of general relativity: the theory of gravity expressed as geometry. Einstein was building a new way to look at gravity as the effect of curved space, employing the mathematics explored in the 1800s by the imaginations of Gauss and Riemann. It was a demanding struggle—by the time Einstein finished his work, he described himself as *"zufrieden aber ziemlich kaputt"* ("content but rather worn out").[1]

Einstein is famous for taking the esthetic approach to physical theory. His innate sense of mathematical beauty helped guide his ideas about how the world works. But no matter how much he joked about instructing the Creator on the proper design of the universe, Einstein knew that the ultimate test of a theory is not how much you like the idea, but how well it describes the real world. In Einstein's curved space, mass (or the mass equivalent of energy) warps the fabric of space–time. Light or physical objects move through that curved space along paths that are determined by the curvature. This was a radical and new approach to gravitation. Einstein knew it was beautiful, but needed experimental tests to see if it was correct.

Albert Einstein diligently computed the orbit of the innermost planet, Mercury, in his new theory. Since Mercury orbits closest to the sun, it feels the strongest gravitational effects, and its orbit was the best place to look for a difference between the new theory and Newton's durable creation of the 1600s. Mercury's orbit is very nearly an ellipse, tracing out the same path around the sun every 88 days. But not quite. The orbit is not exactly closed, so like a giant spirograph, the long axis of Mercury's orbit slowly swings around, advancing 565 arcseconds[2] per century, so that the direction the long axis of the orbit points will make a complete circuit in 225,000 years. In Newtonian gravity, this "precession," the slow reorientation of the orbit in space, is caused by the gravitational effects of the other planets, most importantly, the most massive—Jupiter. In 1859, Leverrier computed the expected amount of precession, later revised by Simon Newcomb as about 43 arcseconds per century smaller than the observed amount. No one understood where this additional precession came from.

One way to get the orbit to rotate slowly with no change to Newton's gravitation would be to have an unseen planet, with the proposed name of Vulcan, close to the sun, hidden from our view, supplying just what was needed to distort the orbit of Mercury. This seems a little far-fetched, because there was no other evidence for Vulcan, though we have grown accustomed to inferring the presence of invisible masses from their observed effects. In fact, there was a strong precedent, since the discovery of the planet Neptune in 1846 followed an analysis of otherwise unexplained motions of the planet Uranus. But in Einstein's theory of gravity the curvature of space near the sun produces just a tiny bit more bending in the path of a planet than you'd calculate from the inverse square law of Newtonian gravity. The net result is just a tiny bit *more* gravitation, a more sharply curved orbit near the sun, and extra precession, without inventing any planets. When Einstein did the arithmetic, he reported feeling "palpitations of the heart."[3] The extra shift in the orbit he computed due to general relativity came out to be 43 arcseconds per century, just what was missing. Quantitative agreement with the facts has the ring of truth. And it is very exciting.

A second test for general relativity was to measure the bending of light as it passed through the warped space near the mass of the sun. This was a more important test than solving the problem with the orbit of Mercury. The discrepancy in Mercury's orbit had been an astronomical riddle for 50 years. The new test was more significant because the same theory, without any adjustments, also predicted a completely new effect that had never been observed. Accounting for the old is good, but making new predictions is an excellent feature for a scientific theory. It gives the observers a way to see if you are wrong. Predictions are a theory's way of living dangerously.

After a false start, Einstein's completed theory predicted a deflection of starlight at the limb of the sun of 1.75 arcseconds, a small but measurable amount. World War I was raging, so even benign communications between Berlin and London were not good. Einstein sent a copy of his paper to Willem de Sitter in Leiden, in the Netherlands, and de Sitter passed on his copy to Arthur Stanley

Eddington in England. In 1916, Eddington was 34 years old, already Plumian Professor of Astronomy at Cambridge and a brilliant theoretical worker who quickly mastered the mathematics of differential geometry that Einstein had employed to describe curved space. Eddington was also in charge of the Royal Astronomical Society's journal, *Monthly Notices*, and he arranged for de Sitter to write three long articles in English that introduced Einstein's new theory to the scientific world outside Germany. Eddington became a powerful champion of Einstein's ideas, promoting them among scientists and explaining them to a wider public.

There is no higher compliment a scientist can give to a theory than personal action to test it. Eddington put his own effort into testing Einstein's prediction. When World War I was concluded by the Armistice in November 1918, Eddington was ready to travel to the island of Principe in the Gulf of Guinea off the coast of Africa for the eclipse of 29 May 1919 while a second expedition traveled to Sobral, in Brazil. By the greatest good fortune, the black sun at the moment of total eclipse would be right in the middle of the Hyades, a group of bright stars that make up the head of Taurus, the Bull. Their undeflected positions could be precisely measured in advance and their positions on the sky should be measurably altered by the warping of space near the sun's edge.

In the aftermath of the First World War, with Berlin still under blockade, this expedition was a touching example of the way science, and especially astronomy, is sometimes able to transcend nationalism. The Earth does look small when viewed from a cosmic perspective, and it is hard to imagine how the energetic fratricide of mustard gas, artillery bombardment, tanks, and trench warfare would look to puzzled observers from Sirius. In any case, Eddington (who was a Quaker and a pacifist) got on a boat to travel for six months to test the predictions of Einstein (who was a pacifist, but definitely not a Quaker). Eddington later called the eclipse measurement "the greatest moment in my life."[4]

The result of this observation, "a deflection of light takes place in the neighborhood of the sun and . . . it is of the amount demanded by Einstein's generalized theory of relativity," was reported to a joint meeting of the Royal Society and the Royal Astronomical

Society, on 6 November 1919 by the Astronomer Royal, Sir Frank Dyson, who had proposed the eclipse expedition. The next morning, the *Times* of London asserted, "it is confidently believed by the greatest experts that enough has been done to overthrow the certainty of ages, and to require a new philosophy of the universe." On hearing of the result, Einstein is reported to have said that if the prediction had not been verified, "Then I would be sorry for the dear Lord—the theory *is* correct."[5]

Observation of new effects that were not predicted in Newton's theory of gravity gave Einstein's radical view of gravity as geometry the weight of truth. Dyson, who reported the measurement, wrote to George Ellery Hale, the creator of the Mount Wilson Observatory in Pasadena, California saying, "I was myself a skeptic, and expected a different result." Hale wrote back disarmingly, "I congratulate you on the splendid results you have obtained though I confess the complications of the theory of relativity are altogether too much for my comprehension. . . . However, this does not decrease my interest in the problem, to which we will try to contribute to the best of our ability." Hale's unfamiliarity with general relativity's rarefied mathematical heights was shared by most astronomers, but his observatory did indeed contribute to the understanding of Einstein's theory, especially as it applies to the universe as a whole. It was at Hale's Mount Wilson Observatory, in the decade after Einstein found himself so suddenly famous, that Edwin Hubble discovered the expansion of the universe. You don't always have to understand the details of the mathematics to contribute to the advance of science. You just have to face in the right direction and go forward with the things that you know how to do.

The bending of light caused by the gravitational field of the sun that Eddington measured in 1919 is small, the measurements were difficult, and, in hindsight, faith in the outcome could have played a part in drawing strong conclusions from uncertain data. But there is no doubt now that the phenomenon is real and independent of the observer's mental state. Gravitational bending of light has been observed in many other settings where it produces dramatic effects that are easy to see with modern equipment. Einstein also predicted, in 1936, that the gravitational field of a star

Figure 4.2. **Gravitational lensing by the galaxy cluster Abell 2218.** The curved arcs are gravitationally lensed images of background galaxies, whose light is bent by the matter (mostly dark) in this cluster of galaxies. Courtesy of NASA, A. Fruchter and the ERO Team (STScI, ST-ECF).

could, in the right circumstances, act like a lens to magnify a background source of light.

In special cases, the immense mass of a galaxy cluster warps the space and acts as a natural lens to make a cosmic magnifying glass. A dense cluster sometimes shows thin arcs around the cluster center. This is not light from galaxies in the cluster, but a mirage caused by mass in the cluster, which distorts the image of yet more distant galaxies. It is a little like looking through the base of a wine glass—distant lights are warped into rings. Gravitational lenses are particularly vivid illustrations of Einstein's idea that mass curves space. They also hint at matter whose effects are important but that is not seen. The light of galaxies is emitted from the hot surfaces of stars, but not all matter is hot and not all matter is in stars. The mass in clusters of galaxies is, for the most part, *not* in the galaxies, but in cold dark matter that we do not see. What is even more peculiar is that most of this dark matter is probably not made of the neutrons, protons, and electrons that constitute our bodies and the world we know. But the lensing effect gives no hint of composition: it depends only on the mass.

Einstein's initial formulation of general relativity, when applied to the universe as a whole, could accommodate either an expanding or contracting universe. Einstein consulted the fog-bound astronomers of 1917. De Sitter correctly reported that the velocities of stars in the grindstone "universe" of the Milky Way were small and gave no hint of cosmic expansion or cosmic contraction. Although his equations looked nicer without it, Einstein faced the facts by sticking an extra term into his equations, the cosmological constant Λ. This created a mathematical solution that Einstein thought made the universe eternal and static (this was later shown not to be quite correct: it could be static, but only for a moment). The cosmological constant is represented in general relativity by the Greek letter lambda. Einstein used lowercase λ but (to make it seem more important in an age of grade inflation) we now use the uppercase Λ. Lambda had no effect on the tests of general relativity in the solar system, but it provided an expansive tendency to space that Einstein adjusted to produce a static universe (if the Milky Way was the universe), as observed.

This mathematical device was completely consistent with his earlier formulation of general relativity, but not necessary. The constant was "cosmological" in the sense that it would make no difference to local physical effects that could be tested by observation in the solar system, such as gravitational bending of light rays by the sun or the advance of the perihelion of Mercury, but would be important only on the largest distance scales. Theoretical physics values simplicity and elegance, and avoids adding mathematical terms that are not compulsory. In fact, this esthetic principle is elevated to a credo—we call it Occam's razor, a pledge to shave ideas down to their essentials. Occam's razor says, "Entities are not to be multiplied without necessity," or more tersely, "simple pictures are best." But Einstein chose to include the cosmological constant. He stuck it in to match the astronomical data.

Einstein apologized for the cosmological term even as he introduced it:

> We admittedly had to introduce an extension of the field equations of gravitation which is not justified by our actual knowledge of gravitation. . . . [The cosmological] term is necessary only for the purpose of making possible a quasi-static distribution of matter, as required by the fact of the small velocities of the stars.[6]

Einstein included the cosmological constant to satisfy the observational evidence as he understood it in 1917. But the observational picture was about to change, and the cosmological constant, already repulsive in one way, was about to acquire a much worse smell. In 1917 astronomers thought the Milky Way was the universe and the velocities of stars were the test for cosmic expansion. But spectra of the "spiral nebulae" and measurements made at the telescopes of Mount Wilson changed all that and turned the cosmological constant into a source of regret.

5

cosmic expansion

Stars were the main business of astronomy in the early 1900s, but a few quirky investigators were trying to understand the spiral nebulae, which look like little pinwheels on astronomical photographs. In the early decades of the twentieth century, Vesto Melvin Slipher worked at the Lowell Observatory in Arizona, a facility established by Percival Lowell. Lowell, scion of Boston industrialists, was fascinated with the idea of studying life on Mars. He used the vast wealth spun out of dark satanic mills on the Merrimack River in Lowell, Massachusetts to build his own observatory near Flagstaff, Arizona, to see what the Martian civilization was up to. Although this sounds as if Lowell was a man whose imagination was running wild, in the late 1800s there was serious discussion of intelligent life with an advanced civilization actively cultivating the planet Mars. Now that we've sent TV cameras, chemistry labs, and gamma-ray spectrometers to the surface of Mars there's less room for speculation. Though there are intriguing signs of water erosion on that planet, and possible microscopic structures in Martian rocks that look like living things, there is no trace of the system of irrigation canals that Lowell wanted to inspect. Instead, Mars looks like Tucson before the developers arrived.[1]

At the time when Einstein was formulating his theory of gravity, the spiral nebulae were thought to be part of our own Milky Way system, and perhaps solar systems in formation, so studying spirals was a reasonable part of the Lowell Observatory's mission. In the new spirit of astrophysics, Slipher undertook heroic efforts with his small telescopes and inefficient photographic plates to obtain spectra of these spiral nebulae. In 1912, he succeeded in getting a spectrum of M31, the Andromeda nebula, and then worked diligently to compile spectra of several more of these enigmatic objects. His spectra of some spiral nebulae resembled the spectra of stars, with the same absorption lines that mark the spectrum of the sun. This identification allowed Slipher to measure the velocity of each nebula from the shift in its spectrum lines. Except for M31 and its satellite M32, almost all the spirals he measured were moving away from us, and many were moving at velocities that were much higher than had been measured for any Milky Way star. Slipher may have thought his measurements were part of learning whether the spiral nebulae are little solar systems in formation. But Arthur Stanley Eddington thought the velocities of the spiral nebulae might be a central clue to cosmology based on general relativity, and included Slipher's as yet unpublished velocities for 41 galaxies, 36 of which were recession velocities, and the largest of which was 1800 kilometers per second, in his 1923 textbook *The Mathematical Theory of Relativity. Somebody* was thinking about the spiral nebulae in connection with general relativity and the possible expansion of the universe! As Eddington put it, "The great preponderance of positive (receding) velocities is very striking."[2]

A galaxy spectrum exhibits familiar absorption or emission lines at an unfamiliar location, shifted toward longer, redder wavelengths. Slipher's heroic collection of 41 galaxy spectra provided half the key to understanding the nature of the expanding universe. The other half came from work by Henrietta Leavitt at the Harvard College Observatory. In Harvard's hierarchical, patriarchic system, the director assigned tasks—and a remarkable group of women carried them out. Harvard had a station in the southern hemisphere at Arequipa, Peru, and it produced a formidable stack of photographs of the Magellanic Clouds to be measured. Henrietta Swan Leavitt sifted through these plates to find the variable stars in the Magellanic Clouds. Scrupulous comparison of one night's data with the next showed that there were many bright variable stars in the

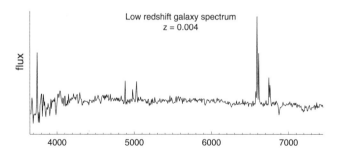

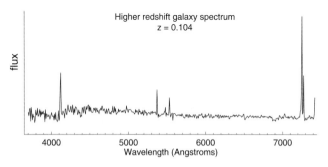

Figure 5.1. **Galaxy redshifts.** The redshift of a galaxy can be measured from the change in wavelength of emission or absorption lines in its spectrum. Cosmic expansion stretches the entire spectrum to the red. Here are two galaxies, one at low redshift, and another at a higher redshift. The spectra are similar, just stretched to the red. Courtesy of Barbara Carter, Harvard-Smithsonian Center for Astrophysics.

Magellanic Clouds, rhythmically growing brighter and dimmer in a regular, periodic way. This type of variable was known from work in our own galaxy: they are called cepheid variables. Cepheids are yellow giants that pulse with periods ranging from a few days to a few months.

In our own galaxy, some cepheids are nearby and some are far away, so it is hard to know the true brightness of a star without some other piece of evidence. A flashlight shining in your eye appears brighter than a lighthouse, or even brighter than a supernova—but it's just closer. In the Magellanic Clouds, the whole system is far enough away from us that all the stars in the cloud are very nearly at the same distance. This means that objects that appear bright really are bright and objects that appear dim are truly intrinsically dim. Henrietta Leavitt used this simple fact to learn something very useful about cepheids.

By 1908, Leavitt found that "the brighter variables have the longer periods."[3] The bright cepheids are physically larger, and their vibrations take longer, much as a big bell sounds a deep note, while dimmer cepheids are smaller, and have quicker pulsations, like a small bell ringing a higher note. This relation between the period and the luminosity was like being able to read the label on a distant light bulb.

You could tell which were the stellar equivalent of 100-watt lamps and which were only 40 watts by measuring something that did not depend on the distance: the period of vibration. The stars were bright (a cepheid with a 30-day period is about 10,000 times as bright as the sun) and the periods were in the convenient range from days to weeks, so cepheids became very useful for gauging the distances of stellar systems. Suppose you found a cepheid in a spiral nebula. If it had the same period as one in the Large Magellanic Cloud, then it presumably had the same intrinsic brightness. By measuring the apparent brightness and applying the inverse square law, you could figure out the distance to the spiral. That would tell whether they were in the Milky Way or not. But in 1920, nobody had done that yet.

In 1920, the National Academy of Sciences sponsored a debate on the nature of the spiral nebulae. Heber D. Curtis argued that the spirals were distant and not part of our Milky Way system. Harlow Shapley, from Mount Wilson, argued against this "island universe" hypothesis. He asserted that the evidence favored the spiral nebulae being part of the Milky Way galaxy. One of Shapley's best arguments concerned the sudden eruption of stars in some of the best-studied spirals. For example, on 20 August 1885, Hartwig at the Dorpat Observatory in Estonia reported a bright new star in the center of M31 that reached 6th magnitude, bright enough to see with a small pair of binoculars. Other novae had been sighted in spiral nebulae. Shapley argued, quite sensibly, that if these stars were like the novae that had been spotted in the nearby regions of the Milky Way, since they appeared to be so bright, it meant the spirals must be nearby and part of our own galaxy.

Otherwise, Shapley noted, if the spiral nebulae were outside the Milky Way, these new stars would have to be ridiculously bright, 100 million times brighter than the sun. It would offend Occam's razor to imagine that there were more types of novae than required by present knowledge. Shapley couldn't imagine "super" novae and considered this "out of the question." Good rhetoric. But not necessarily good science.

On the other side, Curtis advanced a number of reasons why the nebulae might be outside the Milky Way, and he demurely countered the problem of the bright novae by saying, "the dispersion of novae in spirals and in our galaxy may reach [a factor of 10,000] . . . a division into two classes is not impossible."[4]

Scientific debates are a sure sign that the data are just not good enough. In other fields, debates or adversary proceedings like a trial may be the best way to find what we will accept as the truth, or at least a verdict. In scientific research, there's a debate only when there isn't decisive evidence, so that a healthy dose of opinion is required to make sense of the available facts. The truth is out there, all right, but we don't yet grasp it. Since the truth is patiently waiting for us to cast off moss-covered errors and illusions, fallible humans have time to blunder their way forward to the real story. The right tools help.

Harlow Shapley left Mount Wilson to become the director of the Harvard College Observatory. During his long and vigorous career, Shapley had a famous round desk, with wedges reserved for observatory business, scientific research, current correspondence, and manuscripts, and he would rotate the appropriate segment for each topic before him during a working day. He had long since retired when I met him in 1970, a small bent man, 85 years old, in a blue suit. The occasion was the tour of inspection by Harvard's Board of Overseers' Visiting Committee, a distinguished group of outsiders who come every few years to take the temperature of Harvard's astronomy department.

As an undergraduate at Harvard, I did a junior project on the Crab Nebula, the remnant of a supernova in our galaxy observed in A.D 1054. That had been fun, though at the time I had no idea I could contribute anything to this field. As a college senior, I worked on ultraviolet observations of the sun, using data from a satellite project led by Leo Goldberg, director of the Harvard College Observatory. Leo was also chairman of the astronomy department and every year he sent around a deftly worded note to all the students. He encouraged us to submit our senior thesis work for something called the Bowdoin Prize.

"The Prize Committee deplore the continuing paucity of entries in the natural sciences."

After I looked up "paucity," I entered my thesis on ultraviolet observations of the sun. Careful inquiry made to the plural committee revealed that, though the prize essay had a strict word limit, pictures (proverbs to the contrary) did not count! I amplified my prose with many illustrations and nudged out entries on symbolism in Joyce to win a prize for "useful and polite literature (in the English language)." Since then I have tried to be both useful and polite. But more the one than the other.

I picked up the prize check on the ninth floor of Harvard's administration building, rode the elevator down to the fourth floor, and endorsed it to pay back a student loan. I still recall the sensation of feeling light as the elevator accelerated downward, and leaden as it stopped. Years later, my mother said, "You should have bought an oriental rug."

As a senior who had written a prize-winning essay, I was trotted out as part of the dog-and-pony show we presented for the nabobs of the visiting committee. As a reward, I was invited to the lunch the observatory had catered. As at other ceremonial occasions, the central participants sat together, while the less significant sat on the periphery. Shapley was seated next to me, in the outermost circle. I wanted to ask him about his discovery of our place in the galaxy and his memory of the debate with Curtis. Alas, he was not interested in anything but his shrimp cocktail, and that small dish took all his attention. Still, it is good to touch the past. After all, Shapley knew George Ellery Hale, and that's the main line of apostolic succession all the way back to Galileo.

Hale's way of "contributing to the best of our ability" to solution

Figure 5.2. **The 100-inch telescope at Mount Wilson.** This telescope was the largest in the world for thirty years after it went into operation in November 1917. Edwin Hubble used the 100-inch to find and measure cepheids in nearby spirals and to obtain galaxy redshifts. Although Mount Wilson is no longer a dark site, the telescope is still in use. Courtesy of The Observatories of the Carnegie Institution of Washington.

of the problems raised by Einstein's difficult theory was a practical one. He built the 100-inch telescope at Mount Wilson, near Pasadena, California. It was the largest telescope in the world from its completion after World War I to the construction of the 200-inch telescope at Palomar after World War II. Mount Wilson is a wonderful site with clear nights and steady air that Hale had been developing for astronomy since 1904. The 100-inch telescope was built in the engineering style of the Titanic—iron and rivets and big electrical switches and snapping, sparking relays that evoke the most stimulating moments of Frankenstein movies in the middle of a quiet observing night. Having prudently avoided all contact with icebergs, the 100-inch, unlike the Titanic, is still in use. However, the relentless growth of the little village of Los Angeles, which had a population of about 150,000 when Mount Wilson was established for astronomy, has made the skies today much too bright for studying faint objects with this telescope.

In the 1920s, this telescope was precisely the right tool to end the debate about distances to the spiral nebulae. And Edwin Powell Hubble, one-time Missouri lawyer, sometime boxer, Rhodes Scholar, artillery captain, Anglophile, pipe smoker, fly fisherman, and agile social climber, was exactly the right person in the right place at the right time to find the decisive data. Hubble worked at the Mount Wilson Observatory at 813 Santa Barbara Street in Pasadena. He used the 100-inch telescope to look for variable stars in spiral nebulae. He found them.

By repeatedly photographing NGC 6822, M33, and M31 and assiduously comparing one image to the next, just as Henrietta Leavitt had done for the Magellanic Clouds, Hubble identified cepheid variable stars in these systems. The cepheids in M31 were about 100 times fainter than the cepheid stars with the same period that Henrietta Leavitt had seen in the Magellanic Clouds. The apparent brightness of a star declines as the inverse square of the distance. For the same stars to be 1/100 as bright, the cepheids in M31 had to be about 10 times as far away. A present-day distance of 165,000 light-years to the Large Magellanic Cloud puts M31 nearly 2 million light-years away. Hubble's discovery of cepheids in these galaxies, reported in the period 1925–1929, showed that these stellar systems were definitely not a solar system in formation as Slipher had surmised, or some odd swirl at the outskirts of our own Milky Way as Shapley had argued. The Andromeda nebula and, by extension, the other spirals were immense and remote stellar systems—galaxies, which are equivalent to the entire Milky Way.

There isn't just one big central galaxy, with us in it and a void around. Luminous stuff in the universe is made up of galaxies, large and small but on the scale of a billion suns, separated by millions of light-years. Hubble also drew attention to the presence of bright novae—like the 1885 event in M31—in remote systems, as an example of "that mysterious class of exceptional novae which attain luminosities that are respectable fractions of the total luminosities of the systems in which they appear."[5] If the galaxies were distant, these were no ordinary novae. These were the objects that Zwicky and Baade would later call the supernovae. Curtis had been right—a

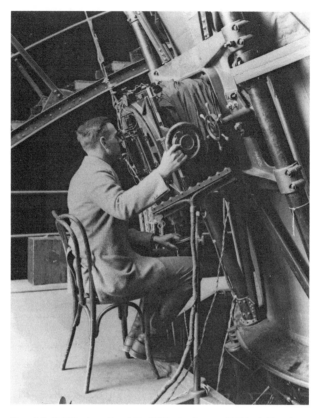

Figure 5.3. **Hubble observing at the 100-inch telescope**. Hubble, clad in jodhpurs and wearing cavalry boots, is perched on a bentwood chair at the Newtonian focus of the 100-inch telescope in 1923. He is holding the controls of the plateholder, which needed constant guiding during the exposure of a photographic plate to compensate for small errors in the telescope drive mechanism. Courtesy of The Observatories of the Carnegie Institution of Washington.

division of the novae into two classes was not impossible. In fact, it was required: there were ordinary novae seen in our galaxy and in the nearest spirals, and much brighter objects, the supernovae, erupting in the distant spiral nebulae.

V. M. Slipher measured velocities of galaxies from the shift of the absorption lines in their spectra. Hubble measured distances to a handful of galaxies using cepheids, then used those to calibrate

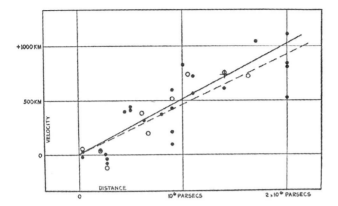

Figure 5.4. **The very first Hubble diagram**. In 1929, Edwin Hubble plotted the velocities of galaxies, determined from their redshifts, against their distances, measured from cepheids and other methods. This diagram shows that the velocity is proportional to the distance, although individual galaxies depart noticeably from this relation and a few very nearby galaxies (like M31) are approaching us. The slope of the Hubble diagram is the Hubble constant, measured in kilometers per second per megaparsec. Hubble's original work showed a slope of 528 kilometers per second per megaparsec, over seven times larger than the modern value near 70 kilometers per second per megaparsec. Courtesy of Publications of the National Academy of Sciences.

the brightest stars in galaxies. The next rung of his ladder of distances resorted to properties of the galaxies themselves to judge still larger distances. The precision of this chain of reasoning was not great, but the early results, though riddled with errors, were enough to show something very profound about the universe.

When scientists have two lists of things—a list of redshifts and a list of distances—you know what they will do. They will plot a graph. That's because we seek the mathematical relation that underpins the observations. The book of nature is written in the language of mathematics, and a graph is the easiest way to see how two quantities are related.

As plotted by Hubble in 1929, the relation between redshift and distance shows that we live in an expanding universe. As Eddington had astutely noted from very fragmentary data 6 years earlier, almost all galaxies are redshifted—moving away from us—and Hubble showed the velocity is proportional to the distance. When you observe a galaxy that is twice as far, you find it is moving away twice as fast. A simple equation connects the measured velocity with the measured distance:

$$\text{Velocity} = (\text{Some number}) \times \text{Distance}$$
$$V = H_\mathrm{o} \times D$$

We call that equation Hubble's law, and the number, the Hubble constant, is the slope of the line in a Hubble diagram of velocity versus distance. We use the symbol H_o for the present-day value of the Hubble constant. The H is for "Hubble" (though he modestly used K, a usage I would like to bring back). H_o is pronounced "aitch-nought" where the "nought" means the Hubble constant measured here and now in the nearby universe. Despite its name, the Hubble constant was different in the distant cosmic past. H_o is measured in astronomers' units of kilometers per second per megaparsec, where a megaparsec is about 3 million light-years. This peculiar form of units keeps the physicists at a respectful distance to avoid contamination.[6]

Hubble's law is definitely *not* common sense—but it is the essential observation that shows we live in an expanding universe. Most of the undergraduates (and I suppose most of the faculty) at my institution are quite self-centered. If they think of Hubble's law at all, they think it confirms their belief that the universe is organized with themselves at the center and everyone else moving away from them. This is the egocentric universe.

But if there is any lesson to be learned from our location in the universe, or any lesson from the history of astronomy, it's that we humans are probably not the central pivot of the universe. The Earth isn't the center of the solar system, the sun isn't the center of our galaxy, and we would be slow learners to insist that our galaxy occupies the central position in the universe.

Instead of assuming that we are at a special place with a unique view of the universe, astronomy today takes the opposite approach. We assume our view is completely typical and the general layout of the universe as viewed from any other location would be the same. To get started, we assume that the universe is the same in all directions and the same from place to place. Of course, we know that isn't true in every detail. All galaxies are not identical, so the view from M31 can't be exactly the same as the view from our galaxy. But if you take a large enough piece of the universe, on average, one region is like another.

Now this is a simple and appealing assumption, but it is also subject to observational test. Unlike political theory, we don't hold scientific truths to be self-evident. We test them by measurement. We can see whether one volume in the universe is like another by making maps of the locations of galaxies and determining empirically how big a patch you need to measure to get a fair sample of the universe. Measuring galaxy redshifts enables you to measure how far away they are, at least for galaxies that are far enough for the cosmic expansion to be larger than the individual motions of galaxies.

In 1983, a group of us glimpsed the biggest structures in the universe from redshift samples of a few hundred galaxies. We were lucky, and detected the Great Void in Boötes, a big region without galaxies about 100 megaparsecs across.[7] Since we only knew this was the biggest structure in our own survey, and it was about the biggest thing we could have seen, we didn't know quite what to make of it. Subsequent redshift surveys led by my colleagues at the Harvard–Smithsonian Center for Astrophysics, Margaret Geller and John Huchra, showed that galaxies form a filamentary structure of great voids and great walls, with features of about the size of the Boötes Void seen in all directions. In the biggest redshift survey of the early 1990s, we showed that once you get to this scale, things seem to even out. We reached the end of greatness and the beginning of homogeneity.[8]

Today, redshift surveys are big enterprises, with redshifts for hundreds of thousands of galaxies being systematically measured by highly automated systems. This field has changed from a cottage industry into assembly-line work. The observed scale of voids and filaments requires that you take a cube of at least a few hundred million light years on a side to get the average properties of the local universe. Once you blur your view to this scale, one piece of the universe is like another. Swiss cheese has a well-determined

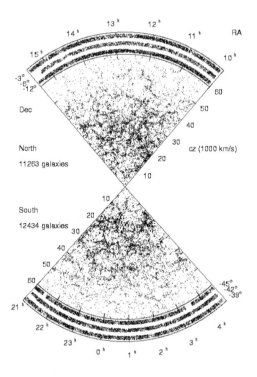

Figure 5.5. **The Las Campanas Redshift Survey.** The redshifts of 23,697 galaxies were measured by a single Harvard graduate student, Huan Lin, as part of this collaboration. The galaxies were selected by their apparent brightness in six thin slices across the sky. This plot, with Las Campanas at the center, uses the redshift and position on the sky to show where the galaxies are located in space. They are clumped, with great voids, great sheets, and great clusters, all on scales less than about 7000 kilometer per second (about 100 megaparsecs for a Hubble constant of 70). On larger scales, the structure seems to even out—this survey was the first that was large enough to see the end of greatness and the beginning of cosmic homogeneity. Courtesy of Huan Lin and the Las Campanas Redshift Survey.

average density once you take a big enough hunk to include both holes and the tasty solids. That's how they can sell it by the pound. A few hundred million light years sounds big for the size of voids, but the observable universe contains over 10,000 cells of this size, so it, too, can have a reasonably well-determined density.

If everyone sees the same universe we do, just from different locations, that's sufficient to make Hubble's law into a recipe for an expanding universe. Start with one dimension—a long, stretchy rubber band. If you glue a little button on the rubber band every centimeter, and then stretch it out, the buttons will move away from one another. If you stretch the rubber band to twice its length, the buttons will each be twice as far apart. If you think about it from the point of view of an ant on each of the buttons, every ant sees all her neighbors moving away, and the more distant ones moving away more rapidly. In fact, this simple stretching produces a displacement, a rate of expansion, which is just proportional to distance. This exactly echoes Hubble's law. It is Hubble's law.

But Hubble's law is not just a demonstration. Hubble's law is measured in the real universe in which we live. The hard part is imagining all of this in two or three or four dimensions. Two dimensions would be something like the stretching surface of a balloon as it is inflated. Ants on the surface would see Hubble's law. In three dimensions, try to imagine a giant jungle gym made of growing bamboo. If you were to hang on to one of the intersections, you'd see all your neighbors receding slowly and distant playmates receding rapidly. You'd see Hubble's law.

When the problem shifts up to three dimensions, it is our own common sense that makes it hard for us to understand these ideas. We can see the weather balloon growing with time, and see the two-dimensional surface stretching. But we're not so good at imagining a space that is expanding in three (or four!) dimensions. A homely, but nourishing, metaphor is to imagine you are a raisin in a baking loaf of raisin bread. As the bread expands in all directions while baking, all the other raisins move away from you, obeying Hubble's law. Cosmic expansion does not depend on an edge and does not need to have a center—to each observer it seems as though the local space is stretching away from you and it seems that you're at the center of your own nutshell.

But that's the same as the view you'd have from any other galaxy. An observer on M31 could invent a common-sense egocentric universe based on observations from M31, another observer from a galaxy in the Virgo Cluster could do the same, and so could an observer in a galaxy deep in a Hubble Space Telescope field. You could say that each of them is equally justified in considering themselves the center of the universe. Which is to say, not at all. Everybody's common sense is just slightly askew because we don't learn the properties of an expanding space from our everyday experiences. Maybe we should plant jungle gyms of live bamboo.

While Doppler's trumpets on a train are a vivid way to see the connection between motion and pitch, the cosmological redshift is not precisely the same thing. It is more helpful to think of the redshift as the effect of the universe stretching out while light travels from a distant galaxy. Light is emitted from a star in a distant galaxy with a particular wavelength set by quantum mechanics. This wavelength gets stretched out by cosmic expansion while the light is in flight. The longer the trip, the greater the redshift. That's Hubble's law. Formally, the redshift is just a number: we use the symbol z for redshift:

$$z = \frac{\text{Wavelength observed}}{\text{Wavelength emitted}} - 1$$

For small redshifts, the speed of light (c) times the redshift (z) gives a velocity. Although we often express redshift as a velocity, it is not exactly a common-sense velocity. The redshift *doesn't* tell us how fast galaxies are moving through a grid of space; it measures the expansion of space that has taken place while the light from a galaxy is in flight to us.

This distinction makes a difference when we measure the velocities of galaxies that are zooming around in clusters of galaxies. There, all the galaxies are essentially at the same distance, and have the same cosmological redshift, but in addition, they have an extra velocity toward or away from us that is due to their own motion in the grip of the local gravitational field. The velocities of galaxies in clusters reveal the amount of matter in the universe by giving a quantitative measure of its gravitational effect. This is how Zwicky first detected the dark matter.

Does the expanding universe means that *everything* around us is growing in size? No.

This is the question correctly answered by Alvy Singer's mother in the first few minutes of Woody Allen's movie *Annie Hall*. Alvy's mother takes him to see the family doctor, Dr. Flicker, because young Alvy, depressed by the meaninglessness of homework in an expanding universe, won't do his. As Alvy explains his angst, his mother interjects, yelling: "What's it to you? Brooklyn is not expanding."

She's right about this. Objects like the Earth (and by extension, Brooklyn) whose structure is determined by electrical repulsion between the electrons in atoms, or by local gravity, do not share the overall expansion of space.

In the 1920s, Einstein's immense prestige, based on the success of general relativity, plus the correct formulation of the problem of an expanding universe by Alexandr Friedmann and others, made the cosmological constant a central element in understanding the universe. The way we usually tell the story, Hubble's 1929 result cut the legs out from under this quest. If the universe is expanding, not static, then there is no need for a cosmological constant. You start with an expanding universe, and it keeps on coasting outward.

By 1931, Einstein had abandoned the cosmological term, noting Hubble's observations "which the theory of general relativity can account for in a natural way, namely, without a lambda term." And he sent it on its way with the curse of sour grapes, saying it was "theoretically unsatisfactory anyway." The legend, promulgated by the physicist George Gamow in his autobiography (but which appears nowhere in Einstein's own writings), is that Einstein called this "perhaps the biggest blunder of my life."[9] I suppose what Einstein (or perhaps Gamow) meant was that if Einstein had ignored the astronomers, stuck with the mathematically beautiful form of his equations, and *not* introduced Λ, he would have *predicted* cosmic expansion a decade ahead of its astronomical discovery, which would have been yet another feat of theoretical brilliance. Of course, he might just as well have predicted cosmic contraction due to gravity, perhaps noting the approach of M31 to the Milky Way, as observed by Slipher in 1912. Then *that* would have been his biggest blunder, revealed when the galaxies beyond M31 did not show blueshifts.

Curiously, Arthur Stanley Eddington, Einstein's great promoter among scientists and envoy to the public, was not so quick to recant. He had noted the evidence of Slipher's velocity measurements in his 1923 book on relativity and he thought the expansion of the universe as observed by Hubble might be the clue to the role of the cosmological constant, which can do more than just balance gravitation: it can cause the expansion and produce an accelerating universe. He did not abandon the cosmological constant in 1929. Eddington explained his ideas in a vivid public talk at the International Astronomical Union's meeting in Cambridge, Massachusetts, in September 1932. He extended the metaphor of the astronomer as sleuth to the breaking point:

> I am a detective in search of a criminal—the cosmological constant. I know he exists, but I do not know his appearance; for instance I do not know if he is a little man or a tall man. . . . The first move was to search for footprints at the scene of the crime. The search has revealed footprints, or what look like footprints—the recession of the spiral nebulae.[10]

Eddington thought the origin of the expansion measured by Hubble might lie in the repulsive effect of the cosmological constant. Perhaps the galaxies had slowly started to expand from rest a very long time ago, and the expansion we see today is just the accumulated effect of Λ accelerating the universe over the eons. So, unlike Einstein, Eddington did *not* abandon Λ, invented to make a static universe, once Hubble had shown the universe was expanding. Instead, he looked to Λ as the source of the observed expansion. If there was an increase in expansion speed over time caused by steady repulsion this would show up in the Hubble diagram, with distant galaxies receding more slowly than you'd expect in a universe that was coasting out from a Big Bang. The measurements of Eddington's time did not extend over large enough distances to look deep into the cosmic past for this accelerating effect. So, although Einstein was done with the cosmological constant, Eddington was not. With a rhetorical flourish that has seemed extravagant, bordering on silly, for most of 60 years, Eddington proclaimed: "If ever the theory of relativity falls into disrepute the cosmical constant

Figure 5.6. **Einstein visits the Mount Wilson Observatory offices.** In 1931, Einstein visited the Pasadena offices of the Mount Wilson Observatory. George Ellery Hale, builder of the 100-inch telescope and founder of the observatory, looks down from his portrait in the library. Hubble (apparently being patted on the head by Hale) is at the left; Einstein, holding chalk, is in front of the blackboard. Courtesy of The Observatories of the Carnegie Institution of Washington.

will be the last stronghold to collapse. To drop the cosmic constant would drop the bottom out of space."[11]

While the theory of relativity has gone from triumph to triumph with the discovery of black holes, images of gravitational lenses, and precision tests of its predictions in the feeble gravity of the solar system and the more powerful tests from neutron stars locked in a close orbit, the cosmological constant acquired a special status as theoretical poison ivy—an idea to be avoided.[12] Eddington wandered farther and farther from the mainstream of theoretical developments in this and in other areas, following his own path into the wilderness.

"The biggest blunder of my life" is Einstein's anathema (whether he said it or not!). From time to time Λ has been picked out of Einstein's trash basket for further examination, but overall, the cos-

Figure 5.7. **The blackboard from Einstein's talk at the Mount Wilson Observatory offices.** This shows Einstein was still using Λ in 1931! Courtesy of the Archives, California Institute of Technology.

mological constant acquired a very bad reputation and was, for the most part, kept out of the discussion of practical cosmology. After all, if it had embarrassed Einstein, what would it do to the rest of us? But, as we will see, the cosmological constant, or something that resembles it very closely, is back again, but this time with evidence. Eddington may yet get the last laugh as we all go diving in Einstein's dumpster.

In 1932, Einstein and de Sitter wrote a paper in which they swore off using the cosmological constant until "an increase in the precision of data derived from observations will enable us in the future to fix its sign and determine its value."[13]

Eddington wasn't ready to give up the cosmological constant, and chided Einstein and de Sitter:

> Einstein came to stay with me shortly afterwards, and I took him to task about [the paper]. He replied: "I did not think the paper [abandoning Λ] important myself, but deSitter was keen on it." Just after Einstein had gone, deSitter wrote to me announcing a visit. He added: "You

will have seen the paper by Einstein and myself. I do not myself consider the result of much importance, but Einstein seemed to think that it was.' "[14]

The application of Einstein's general relativity to the expansion of the universe was worked out in 1922 by the Russian meteorologist Alexandr Friedmann, reinvented by the Belgian Abbé Georges Lemâitre in 1927, and discovered for the third time by physicist Howard P. Robertson. Even before Hubble's discovery, the connection between the expanding universe and gravitation was reasonably well understood. Gravitation slows cosmic expansion.

If, for the moment, you follow Einstein's example (but not Eddington's) after the discovery of cosmic expansion and put the cosmological constant on the shelf, the possibilities are limited. In that case, the expansion of the universe is completely governed by the competition between motion, as expressed by the Hubble constant, H_o, and gravitation, given by the density of gravitating mass–energy. We have a shorthand for talking about the average density of the universe. We compare the observed density with a "critical density" that divides expansion forever from contraction at some far-off time in the future. The ratio of these two densities is just a number: to give it a ring of the ultimate and a whiff of eschatology, we use the last letter of the Greek alphabet, Ω, (omega) as the symbol for that ratio.

The simplest picture is one where there is no matter. Or, anyway, not enough matter to matter. If Ω, the density of matter divided by the critical density, is near zero, and if the universe starts off expanding, later expansion will not be significantly slowed by gravity. Cosmic expansion would continue without limit, neither decelerating nor accelerating, but coasting on indefinitely. If you start with a Big Bang everywhere, you get an expanding universe with Hubble's law for every observer.

If the universe has an appreciable mass density, with Ω_m of 0.3, or 0.6 or 0.9, or any value smaller than one, the universe will still continue to expand without limit as in the $\Omega = 0$ case. Here, I've written Ω_m, where the subscript "m" is to remind us we're discussing the effects of matter, without including the cosmological constant.

Friedmann's solutions for general relativity predict the course of cosmic expansion if you start with an expanding universe. In the presence of significant amounts of gravitating matter, gravitation slows expansion. A universe with Ω_m less than one will grow more dilute as it continues to expand—when you work out the physical details, the expansion will never stop. Even though it is always slowing down, an expanding universe with Ω below one will keep on keeping on, expanding forever.

The critical density itself is the amazingly small number of about 10^{-26} kilograms per cubic meter, or about 6 hydrogen atoms in a typical cubic meter of the universe.[15] Our common sense world of everyday things does not give us a feeling for these numbers. In the room where you're sitting, the air has about 10^{25} particles in every cubic meter. A very good laboratory "vacuum," say in the beam line of particle accelerator or the aluminizing tank at an observatory, might have 10^{15} atoms in a cubic meter. What we think of as "empty" is a million billion times above the cosmic average for the universe. One path to forecasting the future of cosmic expansion would be to take the average number of galaxies per cubic megaparsec from a big sample like the Las Campanas Redshift Survey and multiply by the mass of each galaxy. When we did this, we found Ω_m for matter that clusters with galaxies is about 0.3 ± 0.1.

There is also a simple mapping between density and the geometry of the universe. If Λ is part of the picture, you have to include its effect by computing the mass equivalent of that vacuum energy, which we call Ω_Λ. General relativity is a thoroughly tested theory of gravitation based on Einstein's idea that matter (and energy) curve space. It turns out that $\Omega = \Omega_m + \Omega_\Lambda = 1$ corresponds to flat space, of the type we all learned about in high school where parallel lines don't meet; Ω greater than 1 corresponds to the geometry of a sphere, like the geometry of the surface of the Earth, where lines of longitude, which look parallel at the equator, intersect at the poles. And a low-density universe, with Ω less than 1, has the geometry of a saddle, in which the relations between distances and angles are the opposite from those seen on a sphere.

The geometry of space is not just an abstraction. If there are objects of constant brightness ("standard candles" in astronomical jargon), or objects of constant size ("standard rulers") then astronomers can make measurements to determine the geometry of the universe. In 1961, Allan Sandage, who was Hubble's only student and his heir in carrying forward the program of observational cosmology, wrote a paper in *The Astrophysical Journal* that set out the program to measure the geometry of the universe and to determine its fate by astronomical observation. The article, "The Ability of the 200-Inch Telescope to Discriminate Between Selected World Models," described how the Hale telescope at Palomar Mountain could be used to measure the shape of the universe and to see the deceleration caused by mass in the universe.[16] Sandage showed that the best method was to measure the relation between redshift and distance for objects in an expanding universe. You determine which of the "selected world models" represents the universe we live in by measuring the present expansion rate and the present rate of deceleration from observations. Most of the discussion in Sandage's classic paper is for the case of $\Lambda = 0$. For completeness, Sandage included a brief section near the end of this long paper that shows how to detect a cosmological constant that would produce an accelerating universe, but the discussion for the next 35 years centered on finding just two numbers: the present expansion rate, the Hubble constant, H_0, and the present rate of deceleration, which (for $\Lambda = 0$) gives Ω.[17]

Sandage's program for the Hale telescope was to make a Hubble diagram that extends over a large enough distance so that the cosmological effects of geometry and deceleration would make a measurable difference in the apparent brightness of an object at a given redshift. For the Earth, with a diameter of 12,000 kilometers, the effects of curvature get noticeable when you travel distances of thousands of kilometers. When you fly across the Atlantic, you definitely want the pilot to take curvature into account, flying over Newfoundland on the way to Paris from New York. That's why a globe is so helpful for understanding big distances, even though a flat, foldable road map will do fine for getting lost in Boston. For the universe, the natural time scale is the expansion time, about 14 billion years, and the natural distance scale is 14 billion light-years. So you need to look back several billion light-years for the global effects predicted by general relativity to make a significant difference. Technical difficulties mount as you push out to great distances where cosmology matters: the objects are exceedingly faint, and you are looking at them when they were very young. Just at distances where the effects of cosmology begin to be important, the uncertainties in the measurements begin to grow large.

For decades, Sandage pursued this program at Palomar with the 200-inch telescope, using the brightest galaxies as standard candles, because you can see them halfway across the universe and they seemed to have a small scatter in their intrinsic brightness. A big galaxy has the brightness of 50 billion stars. But galaxies are funny things. They are not really single "things" at all, but collections of stars, and the stars themselves change their brightness as they age over times of a few billion years. Also, galaxies are not so small compared to the separation between them, so in a few billion years, galaxies, especially galaxies in clusters, collide, merge, and grow. These changes in galactic properties can mask the subtle changes in brightness with redshift that cosmology produces. The 25-year enterprise of determining the shape and fate of the universe by observing galaxies did not produce a conclusive result.[18] But applying the same ideas to better-behaved standard candles, type Ia supernovae, with the more powerful telescopes that have superceded the 200-inch has given a strong and unexpected indication of the history of cosmic expansion. The cosmological constant is back: only this time, with evidence.

6

what time is it?

Looking back from the twenty-first century, it is easy to see that Slipher's observations of galaxy redshifts and Hubble's measurements of galaxy distances provided evidence that we live in a large and dynamic universe of galaxies. But in the 1930s, it was not so clear how to interpret galaxy redshifts. The cosmic timescale inferred from stars and the timescale from cosmic expansion did not agree. The connection between cosmic expansion and general relativity had been old business on astronomy's docket since 1917, but Hubble's observations did not reach far enough out in space or far enough back in time to trace the history of cosmic expansion. Direct evidence that the temperature and density in the universe have changed over time is much more recent. Now we can show that cosmic expansion is not an illusion or an assumption, but a fact of cosmic history. The universe has evolved from a hot, dense, and nearly uniform soup of subatomic particles to the present cold, lumpy chowder of voids, clusters, galaxies, stars, and planets made of elements from hydrogen and helium to zinc and uranium. Contrast grew through the accumulated action of gravity, microscopic structure developed through the element-building action of stars, and hot grew cold through the simple chilling fact of cosmic expansion.

If the expanding universe is a real historical account of the past, we can ask some simple quantitative questions. These are the classic traveler's questions, usually asked in a piercing tone from the back seat of the family car: "Where are we?" "What time is it?" and that hardy perennial, "When do we get there?"

We now know where we are. We're little animals on the surface of one small planet, orbiting a middleweight, middle-aged, mediocre star out in one of the spiral arms of the Milky Way galaxy, a flattened disk of 100 billion stars. Our galaxy is just one of 100 billion equivalent systems, a few million light-years from each other, sprinkled through an observable region of 14 billion light-years. Galaxies are clustered into larger swarms of several, dozens, or thousands of galaxies, and there are yawning voids where galaxies are rare for stretches of hundreds of millions of light-years. On scales larger than that, the ups and downs appear to average out and we can speak with confidence about the average properties of the universe. For example, we can measure the average amount of light produced by a cubic megaparsec of the universe from a galaxy census. Once we determine how much mass goes along with the light of those galaxies, we can estimate the density of gravitating matter in the universe, Ω_m, one of the essential ingredients in connecting the contents of the universe with its geometry and its future.

What time is it? How long has the expanding universe been expanding? How old is the world? This interesting question has been approached in many ways. An early effort that predates relativistic cosmology employed Biblical accounts to compute the time since Creation. Adding up the ages of Seth, Enoch, Jared, and Methusela all the way back to Adam, Bishop James Ussher found in 1658 that the time of beginning was at 6 P.M. on the evening of Saturday, 22 October 4004 B.C. By this reckoning, the world began about 6000 years ago. This was a serious attempt to use human history to find the cosmic timescale. Although this earnest effort seems vaguely comic now, Ussher's work shows that the reach of human history and legend goes back just a few thousand years. But we shouldn't be too quick to conclude that the timescale of recorded history is the entire human span or that the human timescale is the cosmic one. The universe is not constructed to a human scale. The age of the world from biblical chronology is just a shake of a lamb's tail compared to the stupefying physical ages of the planets and stars or the ponderous expansion timescale for the universe. It's not even very long compared to the million-year human timescales uncovered by anthropology.[1]

If you concentrate on your earliest memory of the oldest person you ever knew, you can probably reach back to somebody who was born about 100 years ago. And they, in turn, could probably remember something about people who were born a hundred years earlier—my father, born in 1919, remembers a few stragglers of the Grand Army of the Republic, Civil War veterans, marching up Fifth Avenue in parades when he was a boy in the 1920s. And the oldest of those slow marchers, perhaps born in the 1840s, surely knew people who had lived in Thomas Jefferson's lifetime. That takes us all the way back to 1776 and the beginning of the United States in just four steps. So even 6000 years back to Adam and Eve cavorting naked in the Garden of Eden would not be an unthinkably long time—it's just 60 or 100 spans of direct human contact. I don't think it is surprising that the recorded chain of human events goes back a few thousand years or that the problems of Cain and Abel, David and Saul sound just like the problems of people today—lust, jealousy, and overweening pride still exist. But in looking back 6000 years, we've barely begun to peer into the canyon of cosmic time. These people seem just like the ones we know today precisely because 6000 years ago was only yesterday.

The ages of rocks give a different perspective than the Rock of Ages. Fossil evidence shows that the earliest human bones are about 2 million years old. Physical estimates for the age of the Earth show that our hominid ancestors were walking around on rocks that were already aged, even when they were dragging their knuckles on them. Radioactive dating now gives an age for the oldest rocks in the solar system of about 4.6 billion years and this provides a spacious arena for the slow evolution of life on Earth.

Humans are some of the most recent living things on Earth. Our recorded history is just a thin veneer on the surface of time and our unrecorded history, a thousand times longer, is also brief compared to the span of cosmic time. The longest afternoon I can remember was part of a family visit to Washington, D.C., when I was 12. After sitting in a Congressman's chair and touring the Senate, we marched off toward the Washington Monument, 7000 feet to the west. It was August, it was Washington, it was hot and humid, and to tired children, that walk seemed to last an eternity. Or at least the time since the Big Bang. Imagine an immense time line stretching along the Mall in Washington from the Capitol to the Washington Monument. If that distance (around 7000 feet) represents the 14 billion year age of the universe, then the scale is 2 million years to the foot. The Big Bang would be at the dome of the Capitol and the moment when hydrogen cooled enough for the universe to turn transparent would be just two inches to the west. Coming down the capitol steps, you'd be in the era when dark matter began to clump into a rich web of dense troughs into which the baryons would drain. Vigorous formation of the very first stars would change the chemistry of the universe and affect the formation of further structure in unpredictable ways. Galaxies, including ours, began to form about 1 billion years after the Big Bang—in this model just 500 feet from the Capitol, which is still on the Capitol's spacious grounds. The buildup of heavy elements to the level that we see in the sun would have taken place down by the Air & Space Museum. The formation of the sun and planets of our solar system 5 billion years ago would be two-thirds of the way down the National Mall, somewhere near the Smithsonian Castle. The first life on Earth, single-celled and reproducing without sex, appeared about 3 billion years ago within the shadow of the Monument on a late afternoon. Two million years of human prehistory would extend only over the last foot. After 6999 feet of cosmic evolution, all the newsworthy events of the stone age would have to be crammed into that little space. (Throgella lights fire! Throg barbeques mastodon! Throg Jr. invents wheel!) The 6000 years since the invention of writing would occupy only the last 4/100 of an inch, about the thickness of a sheet of cardboard. You'd have to write all of Bishop Ussher's begats on the edge! Deep time of geology, astronomy, and cosmology is not something you can expect to sample through the written record of human culture. The Big History since the Big Bang goes far beyond our common sense or our collective memory.

We can estimate the current age of the universe from the observed rate of expansion. Perhaps an extended metaphor will help.

The city of Boston arranges a big sporting event each year to illustrate cosmic expansion: the Boston Marathon. The wonderful thing about a marathon is that justice is served. In great heaping portions. And quite precisely, because in the long run, details of the start don't matter: once the race is under way in earnest, you don't need Karl Friedrich Gauss to tell you that the runner next to you has the same average speed.

The people ahead are faster; those behind (if any) are slower. For simplicity, imagine a marathon where the runners never tire (isn't theory fun?) so each one runs at a constant speed for the whole 26 miles. Now imagine that a good long time after the starting gun, a runner we shall call Eddie, who departed from the start in Hopkinton with some unusual equipment, starts to make measurements amid the thin-clad throng. Eddie picks out a runner up ahead with green shorts and uses a radar gun to determine her distance and her speed relative to his own. His radar gun, of the type used by police and baseball scouts, emits radio waves with a well-determined wavelength. Those waves travel at the speed of light, bounce off the green shorts of his target, and return. Since his target is ahead of Eddie, it is somebody who is running away from him, and the reflected waves are Doppler shifted to longer wavelength by that motion. The radar gun measures the difference in wavelength between the outgoing radio waves and the reflected radio waves, to compute the speed of the moving target.

Eddie also has a little electronic stopwatch that clocks the time it takes for the radio waves to zip out at the speed of light (1 foot per nanosecond) and return. Surveyors have gadgets like this. So Eddie has measured the speed of the green-clad runner away from him and the distance, too. Suppose he finds that she is moving away from him at 1 mile per hour and that she is 5280 nanoseconds (1 mile) distant. Eddie, who is not really trying to win this race, writes the speed and distance of the emerald-clad strider down in his notebook. Then he turns around and runs backward for a few paces while he records similar measurements for somebody wearing blue shorts 1 mile behind him. Mr. Blue is also receding from Eddie at 1 mph. Now he gets ambitious and measures the distance and speed of somebody in a red outfit 2 miles ahead. Sure enough, Ms. Red got there fair and square by running faster, and the Doppler shift shows she is moving away from Eddie at 2 mph. Same thing for somebody in yellow 2 miles behind—he is also moving away from Eddie at 2 mph. In the stretching pack of runners in a marathon every runner in the race (like an anxious Chairman of the Federal Reserve) sees recession behind and recession ahead. More distant runners are receding more rapidly. This is exactly Hubble's law: the velocity of any runner in the race that Eddie measures is proportional to their distance from Eddie. The stretching out of the runners along the route gives Eddie and each marathon participant the same view that Edwin Hubble discovered looking at other galaxies from our own: nearby objects move away slowly, and distant ones recede rapidly.

Now what's the payoff for slogging through those long paragraphs: a bowl of stew, a Mylar blanket, and a laurel wreath? No. Something much more valuable: a way to measure the age of the universe. What if our mythical Eddie, so wrapped up in his radar gun, his notebook, and his obsession with other peoples' shorts, forgot to bring his wristwatch? With the observations jotted in his notebook, Eddie would have no problem at all deducing how long it had been since the big bang of the start cannon. If he sees somebody running 2 miles an hour faster at a distance of 2 miles, Eddie wouldn't need his watch (or his calculator) to figure the race had been underway for exactly one hour.

Here's the interesting part: he'd compute the *same* time whether he analyzed data for the runner in the green shorts, the one in red, the one in blue, or the one in yellow. One hour. What's more, the view of the race by the runner in yellow, blue, red, or green also would be the same, and any one of them would compute the same time elapsed since the start by observing Eddie or each other. A marathon, which is a one-dimensional mob of runners expanding along the racecourse, has a mathematical relation just like Hubble's law that connects the distance and recession speed of every single runner, and exactly one age that Eddie, you, or any member of the Boston Athletic Association could deduce directly from the same information that goes into the Hubble law.

These ideas also apply in a limited way to the expansion we observe in the universe. Based on the relation between velocity and distance we measure nearby, Hubble's law, you can estimate the time for any galaxy to get to the distance where we see it now. In a homogeneous universe that's the same in all directions, it doesn't matter which galaxy you pick—if you look at one twice as far away, the velocity will be twice as big, and the time will work out to be the same.

Since the distance and the speed increase the same way, the "expansion time" is independent of the distance—it's just

$$\text{Time} = \frac{1}{H_o}$$

Nearby galaxies are receding slowly and distant ones rapidly, but Hubble's law implies there is just one time connected with the expansion, the "Hubble time," which is given by $1/H_o$. A Hubble constant of 70 kilometers per second per megaparsec corresponds to a Hubble time of about 14 billion years.[2]

If the universe has been expanding at a constant rate, then the universe has been expanding for about 14 billion years. If this picture is a real historical account, then there was a time, about 14 billion years ago when the universe was very dense, and everything we see (and don't see) today results from the elaboration of matter and energy since that Big Bang.

A marathon is not a perfect analogy for the expansion from the Big Bang because if the universe were really an explosion from a point, with the galaxies radiating out in all directions, just sorted out by their speeds, the density of galaxies would drop off rapidly as you looked to larger and larger distances. Galaxies do not thin out with distance. The universe is thick with galaxies all the way out. This evidence favors a universe that is homogeneous and isotropic—the same (once you average over big enough regions) everywhere and in all directions. The Big Bang is *not* like an explosion with galaxies shot out as shrapnel. The Big Bang is not centered at a particular location—when we look in any direction, we see distant objects. The Big Bang is the moment when cosmic expansion began throughout the universe.

Did the universe as we know it really begin 14 billion years ago? Common sense (to say nothing of dogmatic belief) balks at the idea of a beginning. Einstein avoided it in 1917 by contriving a static, eternal solution through the device of the cosmological constant. Today the concordance of the Hubble time with independent ways of measuring cosmic ages, physical evidence from the glow left over from the Big Bang, helium synthesized before the first stars, and from direct observation of changes in the contents of the universe over time all point to this as a real chronology for the physical world that could fill up the giant blank spaces on a time line along the National Mall.

Since the Hubble constant is measured by a galaxy's recession speed divided by its distance, our knowledge of the Hubble constant, H_o, and the universal age it implies, $t_o = 1/H_o$, is no better than our knowledge of distances to galaxies. Measuring distances to galaxies was half of Hubble's great contribution, showing clearly that M33 and M31 and the other spiral nebulae were not part of our Milky Way, but the early measurements were not as accurate as the people who made them thought they were. In fact, sharp-eyed hindsight over the last 70 years shows that astronomical measurements are almost never as good as the people doing them think they are. An objective way to estimate the Hubble constant and to determine the quality of that difficult measurement seems to get colored a bit by the affection of scientists for their own results. Of course, this applies only to *other* scientists.

In Hubble's day, the Hubble constant was quoted (with a meretricious air of precision) as 528 kilometers per second per megaparsec, corresponding to an expansion time of 2 billion years. Since the age of the Earth based on radioactive decay was, at that time, estimated to be in the range from 1.6 to 3 billion years (up from briefer estimates based on the time for the sun to radiate away its store of heat), there was the possibility of taking the Hubble time seriously as real evidence for the age of the universe.

Hubble himself was extremely wary about interpreting the redshifts in this way. Perhaps it was the legendary difficulty of understanding general relativity.[3] Perhaps it was the novelty of thinking about an evolving universe, or perhaps it was caution induced as an allergic reaction to that wild man Fritz Zwicky down Lake Avenue at Caltech, who suggested, among many other possibilities, that the redshift might be due not to expansion but to a loss of energy by photons as they traveled through space. Maybe light, like real marathon runners, got tired. In any case, Hubble wasn't too quick to assert that the observed redshift was a genuine measure of expansion, preferring to use the noncommittal term "apparent velocity." And Hubble was also cautious not to conclude that the observed velocity–distance relation implied expansion driven by a cosmological constant of the type de Sitter investigated and which Eddington was eager to see. Hubble wrote a diffident note to de Sitter in 1931, speaking for himself and his observing partner, Milton Humason:

> Mr. Humason and I are both deeply sensible of your gracious appreciation of the papers on velocities and distances of nebulae. We use the word "apparent" velocities in order to emphasize the empirical features of the correlation. The interpretation, we feel, should be left to you and the very few others who are competent to discuss the matter with authority.[4]

Hubble stayed above the fray, but others, competent or not, plunged into the question of inferring the age of the universe from the Hubble law. The results were puzzling. In Hubble's time, the time for cosmic expansion computed from $1/H_0$ was not yet part of a coherent picture for cosmic history. In the 1930s stellar lifetimes were not well understood and they formed the biggest barrier to a concordance of timescales.

In the 1920s, Eddington and other theorists began to see that the energy for stars was connected to the structure of matter on the smallest scales, and might come from subatomic changes. But computing the energy available to a star based on the exchange of *all* its mass for radiant energy through $E = mc^2$ was a vast overestimate. Early workers, aware of the possibility of exchanging energy for mass, but with no way to understand the detailed mechanism before nuclei themselves were understood, assumed that to compute the energy a star can produce, the *m* should be the entire mass of the star. To answer the question, How long could a star shine? they divided the energy calculated from the star's entire mass by the current rate of energy use to get the lifetime. This worked out to thousands of billions of years! Since Hubble's expansion age for the universe was about 2 billion years, this reckoning made the stars in the universe thousands of times *older* than the universe itself. Not a good fit. Astronomers, who thought they knew the ages of stars, were not eager to embrace the implications of a Hubble constant of 528 kilometers per second per megaparsec.

In 1932, Eddington vividly described the problem with revising stellar ages just as the world economy slumped into the Depression:

> Thus astronomers, who have been luxuriating in an enormously long time-scale are threatened with a drastic cut. Even in these days of economy, a cut of about 99 percent is not to be accepted lightly by the department concerned. I confess that I do not quite see how we are going to manage on the reduced allowance; and I am not disposed to blame those whose reaction is to seek for some loophole by which the cut can be avoided.[5]

Acceleration attributed to the cosmological constant can change the relation between the Hubble constant, H_0, and the present age of an expanding universe, t_0. The cosmological constant provides one "loophole" to diminish the problem of the time scales. But in 1932 the real problems were much deeper: the stellar timescale was based on incomplete physics and the astronomical timescale was based on flawed measurements of the distances to galaxies.

In the early 1930s nobody understood enough about the nuclei of atoms to figure out the steps by which stars generate their energy from fusion. One essential missing piece was the neutron—a particle of nuclear physics that was not discovered until 1931. With this neutral partner of the proton in hand, the structure of nuclei, made of neutrons and protons, could be coupled to the problem of cosmic time. Understanding the microscopic world is often the key to unlocking the universe on the largest scales.

Later in the 1930s, with a clear understanding of the possible combinations of neutrons and protons that make up the nuclei of

WHAT TIME IS IT?

the light elements, and encyclopedic information on the nuclear transformations that could yield energy from fusion, Hans Bethe and others puzzled out the chain of nuclear reactions that takes place in the center of the sun to keep it shining. Bethe received the Nobel Prize for this work in 1967.[6] The subtle transformations Bethe identified almost 70 years ago fuse hydrogen into helium, but release less than 1 percent of the mass as energy. In a realistic model, only the center of a star is hot enough for whizzing protons to slam into each other hard enough for nuclear reactions to take place. Most of the mass of a star is left on the sidelines of nuclear burning. As a result of these adjustments, the theoretical lifetime for a star like the sun was pushed down by a factor of 1000 to about 10 billion years. It is understandable that Hubble was guarded in his interpretation of the redshift–distance relation as evidence for real expansion over cosmic time. With the stellar chronometer in such bad repair, it was hard to tell the cosmic time. When the Hubble constant gave a timescale of 2 billion years and the stars gave 10 billion, it was hard to think of astronomical times as a real history of the physical world.

Disagreement over timescales was not entirely the result of theoretical misunderstanding of how stars shine. To make matters more confusing, the value of the Hubble constant derived by Hubble and Humason was seriously inaccurate due to a long chain of subtle errors in measuring distances to galaxies.

As the Irish have many words for rain, and the Inuit many words for snow, astronomers have many words for error. There are at least two ways to get misleading answers from observations. One is to make poor observations with bad tools, so that the uncertainty attached to each measurement is very large. The errors are random, but large. This is often the state of affairs in a new field, like observational cosmology in the 1930s. You are doing your best; it just isn't very good. The other is to make systematic errors so your measurements agree with each other every time, but not with the real value, because you are measuring the wrong thing or you are repeating the same mistake over and over. This is worse, because it is harder to detect. Ferreting out systematic errors requires careful thinking, or, best of all, an independent way to measure the same quantity.

Perhaps a homely example will help. Suppose you are trying to measure the thickness of a single sheet of corrugated cardboard like the ones used to separate the glass sheets in a box of photographic plates Hubble might have employed on Mount Wilson. If you only have a ruler marked in inches, your result for skinny items like these, around a tenth of an inch thick, might be so imprecise as to be nearly useless. Even if you repeated the measurement a hundred times and took the average, you would still have only a vague idea of the right answer because each individual measurement was so crude. Hubble had some random errors of this type, trying to measure the brightness of stars that were right at the limit of what could be detected using photographic plates at the Mount Wilson telescopes.

But far more troublesome are observational errors of another type. Suppose that instead of struggling with a crude ruler, a clever gadgeteer constructs a beautiful micrometer to measure the cardboard's thickness. With a micrometer, you turn a finely calibrated screw until the jaws of an opening just touch both sides of the object you're measuring, and then you read off the thickness on a helical scale. It's the right tool for this job and could give you a precision of 0.001 inch or better. But even with that fine tool, you could make a much more serious systematic measuring error. Suppose you are in the habit of turning that screw just a little too firmly, and without knowing it, you squash the cardboard every time you make a measurement. Then, even though your measurements are quite precise, good to 1/1000 of an inch, they will *all* be too small to tell the real thickness of the sheets, because every time you clamp down the measuring tool, you crunch the cardboard. Your measurements will be precise but not accurate because you systematically measure the wrong thing. You would be making a big mistake to trust the scatter of the measurements to give a true estimate of the uncertainty. You could easily lose your goldfish or your dog from blindly following Gaussian statistics if you have a built-in systematic error.

The opticians fabricating the primary mirror for the Hubble Space Telescope made this kind of error—they were testing the mirror with a lens located very precisely in the wrong place, resulting in a mirror made perfectly to the wrong shape. Hubble him-

Precision

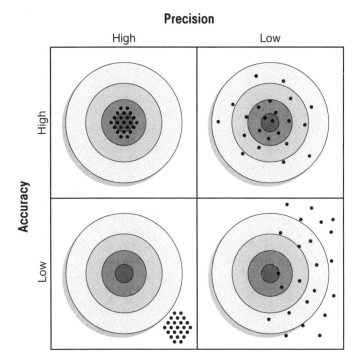

Figure 6.1. **Accuracy and precision.** Accurate measurements have the right average value. Precise measurements are tightly bunched. High accuracy and high precision is best. High accuracy and low precision isn't so great, but it is better than high precision and low accuracy, which conveys a meretricious air of authority to a misleading result.

self had systematic problems in correctly connecting the dots that stars formed on photographic plates with the true brightness of those stars. Just because your measurements agree with one another is not a guarantee you're doing things right.

There are other subtle ways to go wrong that have to do with what's in your sample of objects to measure. Suppose the stack of cardboard has some thin shirt cardboards from a Rhodes Scholar's starched Oxford cloth shirts mixed in with the corrugated sheets from the plate box. Even if you carefully measure a hundred sheets from this stack, correctly compute the average, and conscientiously report it to 1/1000 of an inch, you will still have the wrong value for the corrugated sheets alone because some interlopers crept into the sample. If you don't have a clear enough understanding of the objects, sampling errors sneak in.

Astronomers are especially vulnerable to a tricky sampling error based on picking objects that are bright enough for you to see. It is such a frequent problem is has its own proper name: Malmquist bias. Nearby, you can see bright supernovae and dim ones. But as you look at more and more distant samples, as the dimming caused by the inverse square law limits your ability to detect objects, the only objects you will see are the especially bright ones because the dim ones don't make it over your detection threshold. The average intrinsic brightness of your sample creeps higher and higher as you look at more distant objects. If you're judging distance from apparent brightness, the distance scale gets compressed at large distances. This is very bad. It's like having a tape measure that starts out with tick marks spaced evenly, but which subtly shifts scale as it plays out. You will systematically underestimate the largest distances because you are inadvertently selecting just the brightest supernovae. You become a victim of Malmquist bias.

It's a little like looking at people walking by on the sidewalk from a ground-floor window. If your window stretches down to the floor, you will see tall people and short people, Chihuahuas and Great Danes passing by. But if the windowsill is 6 feet off the ground, you'll miss all the dogs and children and you might conclude that everybody in your town is 6 feet tall. It's just everybody *you can see.* And that's not always the same thing.

And there are ways to goof that are not so subtle. Suppose the fine Swiss micrometer you are using is calibrated in centimeters, but you're a rocket scientist in the United States and you think it is in inches. This type of error is known technically as a scale error or "stupid mistake." Everything you report will be off by a factor of 2.54 even though every measurement looks as if it is good to 0.1 percent. These measurements will be precise, but not accurate and your spacecraft will disappear near Mars instead of landing on it. Usually, the evidence that you are making an error is not so vivid.[7]

Systematic errors are much worse than crude measurements with big measuring uncertainties. Instead of leading to vague conclusions, a measurement that is precise but not accurate can lead to strongly held, but wrong, conclusions. Hubble's measurement of 528 kilometers per second per megaparsec has an air of precision (it's that "28" that makes you think he's pinned it down to just a few kilometers per second per megaparsec), but it was seriously inaccurate because of several subtle systematic errors that crept into his measurements during the night. For example, Hubble identified cepheid variables in nearby galaxies and then compared their brightness to similar cepheids in the Magellanic Clouds. The distances to the Magellanic Clouds were wrong by a factor of 3. This is like having a micrometer that you think is measuring inches, but it's really centimeters. When you use the work of others, if they are wrong, you get the wrong answer too.

The cepheids turned out to be more complex than Hubble knew, and he was lumping together two different types of variable stars when he made his measurements. This is a little like having two kinds of cardboard in the stack. To get out to the Virgo Cluster, at a redshift of about 1200 kilometers per second, Hubble needed something brighter than the cepheids. But it turns out that the brightest "stars" that he picked out from his Mount Wilson plates to measure those distances were not really stars at all. They were giant clouds of gas, glowing because the gas is excited by the ultraviolet light of many massive stars within. This is a little like squashing the cardboard—you're not measuring the quantity you think you are and you do it over and over and over without realizing it. The accumulation of all these systematic errors in Hubble's work is very significant. Modern values for the Hubble constant are seven times smaller than Hubble's and for the Hubble time, seven times larger. While Hubble's Hubble constant corresponded to a disturbingly short Hubble time of 2 billion years, today there is reasonable agreement between the Hubble time and stellar timescales *if* the universe has been expanding at a constant rate. That's a big "if" because it is exceedingly difficult to measure changes in cosmic expansion.

We have good reasons to believe that the errors of measurement for the Hubble constant are smaller today than they were in the 1930s. But systematic errors that depend on the distance to the

Magellanic Clouds, understanding the types of supernovae, the properties of cepheids in different settings, possible confusion of a single star with many, and the dreaded Malmquist bias are still with us. The challenge for observational astronomers is to anticipate possible systematic errors and then try to limit them through measurement. But the human mind is fallible and there are many ways to make subtle, but significant, errors or even mistakes, which are not improved by a refined statistical treatment of the data. Sometimes it's the problem you didn't think of that jumps up to bite you, as with the pulsar in SN 1987A. You can be sure you've got the right answer only when independent lines of evidence converge. Different groups of people and independent ways of making the measurement guard against the many forms of error.

Today there is still a lively discussion on the numerical value of the Hubble constant. The relation between distance and redshift itself is not in doubt, but the numerical value of the slope, which is H_0, has been difficult to measure accurately. Hubble's old value of 528 kilometers per second per megaparsec is so far off the mark, it hasn't been part of the modern discussion. Values since 1950 range from 50 to 100 kilometers per second per megaparsec, with the most recent measurements narrowing down to the range from 60 to 80. In this book, I use 70 ± 7 kilometers per second per megaparsec because I think it well represents today's data, especially the data from supernovae.

The basic techniques that Hubble used in the 1920s are still right at the center of modern measurement. Cepheid variable stars play the leading role in establishing the cosmic expansion rate, just as they did in the era of silent films. What has changed are the tools for measuring light from distant stars.

Galileo led the way in applying telescopes to astronomy. When you go to Florence, you can nip up to the Museum of Science while somebody is holding your place in the long line for the Uffizi Gallery. Galileo's 1610 lens is enshrined there (along with Galileo's own finger, like the relic of a saint who was no saint). With his first astronomical telescope, Galileo used his eye to detect craters on the moon (and measure their height), to see Jupiter's moons orbiting that massive object (as the planets orbit the sun), to observe the phases of Venus (a prediction of the Copernican view of the solar system), and to observe that the band of light across the summer sky, the Milky Way, was not made of Hera's milk as legend held, but of a vast number of stars, too numerous to be resolved individually with the naked eye. This was very good work for a small telescope. Galileo immediately applied for a research grant from the Medici. This explains why modern scientists find Galileo a kindred soul.

One great advance from Galileo's time to Hubble's time was the steady march toward larger telescopes. The drum major in this parade was George Ellery Hale who orchestrated building the world's largest telescope four times: the 40-inch at the University of Chicago's Yerkes Observatory in 1887, the 60-inch reflector at Mount Wilson in 1904, the 100-inch telescope at Mount Wilson in 1917, and the 200-inch reflector at Palomar Mountain, now called the Hale Telescope, which started operation in 1948. Big telescopes collect more light and, other things being equal, enable you to measure fainter and more distant objects.

The other great advance was the invention of better detectors to measure the light that giant telescopes gather at such great expense of money and effort. Galileo's eye was developed by natural selection over the last few hundred million years, and was a marvelous light detector (until he went blind), but eyes are limited in two fundamental ways. First, there's no permanent record—you can have eyewitness accounts, and drawings, but there's no way to store the actual data. Second, you can't accumulate the light in a time exposure to record fainter objects than you can see in a single good look. The apogee of the eyeball approach to astronomy was the "Leviathan of Parsonstown" telescope completed by William Parsons, third Earl of Ross, on his commodious front lawn at Birr Castle in Ireland in 1845. The telescope, with a 6-foot, 3-ton metal mirror, and an ingenious pointing mechanism of chains and cables between massive masonry walls, had elaborate wooden scaffolding to lift the observer to the business end of this monster so he could look in by eye. There's also a nice drawing board so the observer could sketch what he saw, if there came a fine night in County Offaly. There must have been a few clear nights, since Parson's sketch of the Whirlpool galaxy, M51, provided the first evidence for the shape of "spiral nebulae."

Every subsequent large telescope has been built with photography in mind. Starting in 1852 with a daguerreotype of the Moon made with the Great Refractor at the Harvard College Observatory and on into the 1970s, astronomical evidence was recorded on photographic plates of the chemical kind: extra-flat glass coated with a gelatin emulsion that suspends silver salts. Plates could be exposed for long periods of time and when later developed, the silver metal retained a record of the stars and galaxies whose light had fallen on them. The advantages were tremendous—long time exposures, like the heroic early galaxy spectra taken by Slipher over many hours, accumulated light for much longer than a human eye, which sums up light for less than a second. And the record was comprehensive and permanent, so Hubble could go back month after month to photograph M31 and then compare the plates, searching over the whole image for stars that had changed their brightness—the cepheid variables that set the distances to nearby galaxies.

The great recent technical change of modern astronomy has been to shift away from these messy, but simple and cheap, analog chemical imaging devices where light plus darkroom voodoo makes dark dots on a glass plate. Now we have complicated and expensive digital imaging. Light falls on small wafers of silicon carefully held deep inside elaborate cryogenic bottles where ancient photons from distant stars liberate electrons that can be measured with a delicate amplifier and recorded digitally in a computer.

Why is this better? It is better because photographic emulsions detect only about 1 percent of the light photons that fall on a plate. Light from a distant supernova travels across 7 billion light years of intergalactic space, traverses Earth's atmosphere, bounces off the primary mirror of a big telescope and into the camera. In Hubble's time, 99 percent of that light was lost right there, being absorbed in the photographic emulsion without making a grain to be developed. What a waste! Silicon CCD (charge coupled device) detectors, sophisticated siblings of those in digital cameras, detect almost 100 percent of the light. So old telescopes with modern detectors are nearly 100 times more efficient than when they were built.

Until very recently these electronic detectors have had one serious drawback. They have been small. In the 1970s, the silicon arrays were about the size of a dainty fingernail. In contrast, from the 1950s, Kodak glass plates 14 inches on a side were standard issue at the Palomar Schmidt telescope. When I was a graduate student at Caltech, astronomy professor Wal Sargent, who had inherited the Palomar Supernova Search when Fritz Zwicky retired, asked me to fill in while the regular observer, Charlie Kowal, was on vacation. I was eager to go up to Palomar to learn how act like an astronomer. Palomar has traditions and hierarchy passed down from Mount Wilson. At lunch, I was issued a cloth napkin for use during the duration of my observing run. It was clipped with a wooden clothespin that had my name written on it in pencil on one side and somebody else's name on the other side. Allan Sandage had a real napkin ring and sat at the head of the table. It took me a year to move up to my own private clothespin. I still have it.

I learned to handle these monstrous, fragile, thin glass Schmidt plates in the dark, determining which side had the gelatin coating and which was plain glass from a delicate taste test (sticky or slick?). I also learned the hard way not to let the plate slide against my fingertips. Failure to observe this rule was punished by a neat cut on the fingertips just before immersing them and the giant plate into the gentle acid of stop bath. I also learned not to get up early to develop the previous night's plates without disarming my alarm clock. When I came back to the monastery for lunch, there was a loud buzz coming from my room. Allan Sandage was not amused.

Plates were painful for the inept and inefficient for everyone, but they were big, and for some purposes, like searching for supernovae, their ability to cover an area on the sky 1000 times larger than a small chip has been more important than the lost factor of 100 in photon detecting efficiency. That summer, I found supernovae 1971M and 1971N. Alas, the discovery plates were the *only* measurements of these two objects, and I'm sorry to report that finding these two supernovae contributed nothing to increasing our understanding of supernovae. But it did increase my understanding of how to make progress in this field: if you don't follow up your discoveries with further observations, then you don't learn anything

other than the vanished skill of how to conduct yourself in a darkroom. My mother-in-law was proud of me, though, for making the
all-time list of supernova discoverers. She carried a picture of those
supernovae around in her wallet and showed them when other
Smith graduates asked her to admire their grandchildren.

In the last decade technological change has tipped the balance;
now searches with electronic detectors are more effective because
the detectors are bigger. CCDs are fabricated by the same techniques used to make the integrated circuits that are the guts of
modern computers. When light falls on silicon, it liberates electrons
that are stored, shuffled along to a delicate amplifier, and read out
to give a quantitative measurement of the amount of light that
fell on each point in the detector array. The CCD detectors we now
use to search for supernovae halfway across the universe are 6
inches across. Recording just one of these images takes about 288
megabytes of computer memory, compared to about 6 megabytes
for a digital camera you can buy today at Circuit City. Being able
to add or subtract big arrays of digits depends on the improved
capacity of computer disks and memories and processors. Fortunately, bigger CCDs and more capable computers all stem from the
same improvements in the technology of etching silicon. Fundamentally these technical advances, rather than some brilliant insight, have led to a major conceptual change in our view of what
makes up the universe.

Over the last decade one of the most important uses of the Hubble Space Telescope (HST) has been to extend Hubble's work on
cepheid variables. HST has modern detectors plus exquisite imaging from above the atmosphere (now that the original error in
the primary mirror has been corrected with smaller optics that work
something like a set of contact lenses) so it can measure cepheids
in galaxies 25 times further away than M31. From the ground, these
cepheid variable stars get blurred with their neighbors and the glow
of the nighttime sky makes them impossible to find or to measure.
Using the Hubble telescope, astronomers can now do what Hubble
aimed to do: measure individual stars in galaxies that are far enough
away to set the cosmic distance scale. By taking repeated images
of galaxies, observers pinpoint the variable stars, determine their

periods, and then gauge the distance to the galaxy from the apparent brightness of cepheids.

The brightness of cepheids in galaxies studied with HST is about 100,000 times fainter than cepheids with the same period of vibration in the Large Magellanic Cloud. This means they are about 300 times as far—something over 50,000,000 light-years from us. Fifty million light-years sounds like a lot, but the stretching out of the universe is a subtle thing and this is not far enough to make a good measurement of the Hubble constant. At that distance, the expansion velocity is only 1200 kilometers per second. This is just 0.4% of the speed of light, and only a few times bigger than the random velocities of 300 kilometers per second that individual galaxies have just milling around among their neighbors. Where we can measure the cepheids, even with HST, the cosmic expansion velocities are too small to trust. Sometimes the velocity of an individual galaxy will be larger than the local Hubble velocity, and sometimes smaller, because each galaxy has been affected by the tugs and pulls of gravity from its neighbors acting over billions of years.

Variation about the Hubble law means that even if there were no errors of any type in the distance measurements out to 50 million light-years it would still be tough to do a good job of determining the Hubble constant. Because the cosmic expansion velocity at that distance is so small, the Hubble constant, which is the ratio of velocity to distance, must have big uncertainties, and inferring the age of the universe from the rate of cosmic expansion is not reliable without a longer yardstick.

This is the principal reason why progress in measuring the Hubble constant has been so difficult: the best cosmic yardsticks are the cepheid variables, but even with HST they are too faint to measure in galaxies that are moving rapidly with universal expansion: out in the Hubble flow. To measure the Hubble constant, you need a good distance tool that can carry you out far beyond a paltry 50 million light-years to galaxies that are receding at 10 percent of the speed of light—out to 1 or 2 billion light-years. Then a few hundred kilometers per second more or less of an individual galaxy's motion wouldn't matter much. It would be swamped by the outward stretching of cosmic expansion at 30,000 kilometers per second. That would give a good measurement of the Hubble constant, and a reasonable way to test the expansion timescale against other cosmic clocks.

The technical problem in doing this measurement is that the apparent brightness of a distant object drops off as the square of the distance. Since the range you want is roughly 30 times the distance you can reach with cepheids, objects will appear 30 × 30, or roughly 1000, times fainter. So you need something much brighter! Cepheids are already among the brightest stars we know, typically 10,000 times brighter than the sun, and the list of brighter objects is short. But there is one kind of stellar event we know that, for a little while, reaches 4 billion times the luminosity of the sun: a supernova explosion. For a few weeks, these incandescent cosmic catastrophes are bright enough to serve as rulers to measure the size of the universe.

The best candidates for measuring the universe are SN Ia, the type of supernova that comes from a thermonuclear explosion of a white dwarf. They are about 100,000 times brighter than a cepheid variable star, so at the distance of a galaxy like M100 where you need HST to see the cepheids, you can easily measure the light from a supernova with a well-equipped amateur telescope on the ground. Today, the very best Hubble diagrams come from carefully corrected measurements of SN Ia.

Unlike cepheids, however, supernovae don't repeat their cycles. It's a brilliant one-way trip to destruction. Worse, we don't always see the whole rise and fall of a supernova light curve in the first month or so after the explosion. Depending on the strategy and diligence of the searchers, supernovae are often discovered after they peak. To use supernovae to measure the Hubble constant, you need a way to use those tardy measurements. You need to see if there's a standard light curve, so that you can extrapolate from the piece of the light curve that you do see to the part you don't.

In 1989, this problem was attacked and solved by Bruno Leibundgut, in his Ph.D. thesis work at Basel, Switzerland, working with Gustav Tammann. Bruno stitched together the light curves of

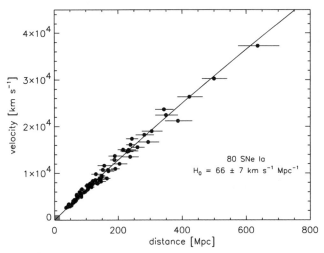

Figure 6.2. **The Hubble diagram for type Ia supernovae.** Note that the velocity is proportional to distance, as noted in 1929. Hubble's original Hubble diagram (figure 5.4) extended only out to 2000 kilometers per second, where the individual motions of galaxies added to the scatter. This Hubble diagram extends out to 30,000 kilometers per second, one-tenth the speed of light, where the cosmological Hubble flow is large compared to any galaxy's individual motion among its neighbors. Courtesy of Adam Riess, Harvard-Smithsonian Center for Astrophysics.

several well-measured SN Ia to construct a template. The assumption underlying Bruno's work was that all the SN Ia were identical. This was plausible in the 1980s: there was a theoretical reason to think exploding white dwarfs at the Chandrasekhar limit were all the same and the observations, mostly photographic, were not good enough to see subtle differences clearly. After finishing his thesis, Bruno came to work with me in Cambridge as a postdoc at the Harvard–Smithsonian Center for Astrophysics. Data assembled by our team and techniques we developed in the 1990s helped make the SN Ia the best standard candles for measuring distances to galaxies.

Bruno had some of the traditional Swiss characteristics: he was very careful, thorough, and self-critical. These are good properties in someone dealing with light curves, where there are many ways to err. But Bruno also had some not-so traditional properties. The joke is that Switzerland doesn't have an army, it *is* an army. Every male is supposed to serve in the army and then keep a rifle (not just a Swiss army knife) at home in good condition for national defense. Einstein was a pacifist during World War I, when he was in Berlin, and a spokesman for world peace even after he helped instigate the Manhattan Project to build a nuclear weapon. But when he was a young man, Einstein was called for the Swiss army— and he was rejected for flat feet and varicose veins. Bruno was also called for the Swiss army, and his feet and veins were fine, but he chose "service without a weapon," as a matter of principle. Bruno is a person who thinks for himself and stands up for his own opinions. These are good properties if you are a Swiss citizen, Gustav Tammann's student, or my postdoc.

For many years, the impeccably dressed, careful, and energetic Tammann had collaborated fruitfully with Allan Sandage. By the late 1980s, they were embarked on a program with HST to discover cepheids in galaxies with well-observed supernovae where Bruno's template was useful. The idea was simple. You make a list of galaxies that have had well-observed supernovae. You use Bruno's template to figure out the peak apparent brightness. Next, select the galaxies that are close enough so that HST can find their cepheids.

Then comes the heavy lifting: you must convince the Space Telescope Time Allocation Committee to let you take many HST images of your target galaxy to find the cepheid variables and to determine their periods and apparent brightnesses. Armed with the apparent brightness of cepheids of known period in a galaxy, you can figure out its distance, just as Hubble did, by comparing those stars to the cepheids in the Large Magellanic Cloud. Now you turn the problem around: if you know the distance to the galaxy and, from Bruno's work, the apparent brightness of the supernova, you can do the arithmetic to see what the intrinsic power output of a supernova is. Do this for enough galaxies to average out the errors (if they are purely Gaussian errors) and you learn the true brightness of SN Ia. Present values are somewhere around 4×10^9 solar luminosities.

The last step is to use a distant set of supernovae found in galaxies out in the Hubble flow where redshift reflects cosmic expansion, not random motions. Use the known luminosity and measured apparent brightness to compute the distance to each one. Divide the velocity (in kilometers per second) by the distance (in megaparsecs), average to decrease the errors, and, *voilà*, you have the Hubble constant.

This program sounds simple, but it isn't. First, supernovae are rare. Since a supernova of this type goes off in a galaxy once every century or so, and we've only been aware of supernovae since Zwicky and Baade's pioneering work in the 1930s, there is only a small number of good cases where a well-observed supernova erupted in a galaxy within HST's limited range to detect the cepheids. Allan Sandage, Gustav Tammann, Abi Saha, and their collaborators, who have been carrying out this program with HST, list 9 supernovae discovered in nearby galaxies. They find a Hubble constant of 60 ± 6 kilometers per second per megaparsec, up from earlier measurements that placed the Hubble constant in the 50s.

Wendy Freedman, three decades younger than Sandage, but also working at the Carnegie Observatories, has been the leader of a "Key Project" team measuring the Hubble constant with HST. Observational cosmology is not exactly a contact sport, but it helps to be tough and competitive if you are working in the same field as Allan Sandage, and at the same institution, and especially if you get a different answer. Wendy and her sister used to play on the University of Toronto's women's hockey team—as an experienced right wing, the rough and tumble of astronomy doesn't bother her too much. Wendy's group has used a number of other methods besides supernovae to measure distances to galaxies. She's looking to see whether they agree, to get around the particular systematic problems of each technique. She calls this "cross-checking" her data, which is bad in hockey, but good in observational astronomy. When the Key Project team first reported their results in 1994, they got relatively high values for H_o—near 80 kilometers per second per megaparsec, which corresponds to a meager allocation of 12 billion years since the Big Bang. Her group's present value of H_o— 72 kilometers per second per megaparsec is widely asserted to be good to 10%. History suggests that we are always confident, but rarely correct. Or, perhaps we really are nearing the end of smoking out errors in the Hubble constant. This is not quite in agreement with Sandage's team, but the differences are getting smaller.[8]

Fundamentally, the accuracy of the Hubble constant, H_o, and the expansion age, t_o, inferred from $1/H_o$ for supernovae depends on distances to nearby galaxies that have hosted SN Ia, which can also be measured with cepheids. The distances to those galaxies depend on comparing their cepheids with the same type of star in the Magellanic Clouds. This step-by-step measurement of the universe leads to an odd situation—our knowledge of the size and age of the entire universe depends on measuring the distance to the nearest galaxy, the LMC. It is frustrating but true that we haven't got this local problem completely solved. The most recent revisions to the Hubble constant depend on improved measurements of cepheids in the Magellanic Clouds, not on supernovae 10,000 times farther away.

How would we know if today's distance scale based on cepheids were wrong? One way is to compare independent methods of measuring distances to the same galaxies. Cepheids underpin most methods for finding the distances to galaxies, yet there is a handful of ways to measure extragalactic distances that doesn't depend on these stars. If independent methods give the same answer, perhaps both methods are measuring the distance accurately. If they disagree, then somebody is wrong.

Supernovae provide two distance-measuring methods that have nothing to do with the distance scale based on cepheids. First is the amazing ring in SN 1987A. When the supernova was first sighted in February 1987 George Sonneborn, at Goddard Space Flight Center, and I compiled a detailed record of the supernova's changes with the International Ultraviolet Explorer satellite. The first change was a little disappointing: the ultraviolet light from the supernova faded very rapidly, plunging by a factor of 1000 in the first three days after discovery. But then, after about 90 days, the ultraviolet spectra began to show something curious: narrow emission lines from highly ionized nitrogen atoms. The fact that SN1987A showed emission lines suggested that this light was coming from gas that had been excited by the supernova explosion. The fact that the lines were narrow meant that the velocity range of the emitting gas was very small. Since the supernova itself was ripped apart by a violent explosion that sent the outer layers flying out at 10% of the speed of light, small speeds for the emitting gas ruled out supernova debris. What was this stuff?

My Swedish colleague, Claes Fransson, had a good idea for something simple that would cover the facts. What if there were a shell of gas around the supernova, perhaps exhaled by the pre-supernova star? If it was located at the right distance, then the powerful flash of light from the explosion would take many months to reach that shell, excite it, and make it glow. A flash of ultraviolet light would rip the electrons off nitrogen atoms to ionize them, but it wouldn't give the gas much of a kick up to high speeds.

If this story was right, then the emission lines we were seeing would get stronger over the next several months. If the shell were big enough, light travel time would matter for us in 1988 just as it did for Ole Rømer in 1676: the near side of the shell would be many light-months closer than the far side. In fact, it took about 400 days for the lines to reach their maximum strength, which we took to mean that the pre-supernova star was surrounded by a spherical shell with a diameter of about 400 light-days.

All of this was very interesting information about the last gasps of a massive star, but it did not yet provide an independent distance to the Large Magellanic Cloud. That came after the 1990 launch of the Hubble Space Telescope. There were hints from early ground-based data that there was something near the supernova that lit up in the months after the explosion. The earliest images from HST, even with its flawed initial vision, showed that supernova 1987A was surrounded by glowing gas, just as Claes had predicted. (See figure 3.3.)

Except, as usual, nature was wilder than our imaginations. The gas was not in a simple shell around the explosion, but in a ring, presumably the inner boundary of a flattened donut of gas. Even with the blurry version of an uncorrected HST picture, you can measure the angular size of the ring. It turns out to be about 1.6 arc-seconds. Since the blurring effect of the Earth's atmosphere is typically around 1 arcsecond at a good astronomical site, HST, sited above the atmosphere, was essential for this measurement. Now, if you know how large the ring is from timing the rise to maximum of the narrow emission lines in the UV, and you figure out the tilt of the ring from its shape or the light curve, and you also know the angle the ring covers as viewed from our galaxy, it is not hard to compute the distance to SN 1987, and hence to the Large Magellanic Cloud.

We found that the distance to the LMC is about 165,000 light-years, which is the same distance that Wendy Freedman and her associates have been using for the beginning of the cepheid-based distance scale. So we agree, using completely independent paths. It could be chance, but it could be we are both right.[9]

We invented another method based on supernovae to check the cepheid distance scale. While SN Ia are thermonuclear explosions in white dwarfs, type II supernovae result from the collapse of a massive star. When the outside of a SN II is ejected, it is still mostly hydrogen. The properties of the expanding, cooling atmosphere can be computed in detail by a very smart graduate student. Ron Eastman, who worked with me at the University of Michigan and at Harvard before joining the scientific staff at Livermore, did this computation in 1989 and worked out a refined method for comparing the models to the data for SN 1987A. Repeated measurements of the temperature, speed, and brightness of the supernova atmosphere supply enough information to figure out how large the atmosphere is, and to compute the distance to the explosion. For SN 1987A, the distance comes out again to be near 165,000 light-years, in good agreement with the conventional distance.[10]

In 1994, for his Ph.D. thesis, my Harvard graduate student Brian Schmidt applied our "expanding photosphere method" to all the available data for explosions since 1969. Interestingly, some of the galaxies with SN II data and expanding photosphere distances are also galaxies in Wendy Freedman's Key Project sample. The results agree very well, which suggests either we are both doing something wrong or both doing something right. Since the two approaches

WHAT TIME IS IT?   111

are completely independent, we suspect this is a clue that we are doing something right.

More than 20 years after I gave a talk to our Visiting Committee and had a dull lunch with the aged Harlow Shapley, I was the department chair at Harvard, trying to set up the program for another Visiting Committee. Wallace Sargent from Caltech was the chair of the Visiting Committee. Since I had been a graduate student at Caltech, Wal knew me pretty well, and I was eager to show that there were good things happening at Harvard. Brian Schmidt seemed the natural choice to talk to this outside group. He had an independent measurement of the Hubble constant, which was a hot topic. Plus, Brian is a charming guy, a lively speaker, and had proved to be a very good teaching assistant. If he could deal with Harvard undergraduates, I reasoned he could deal with Caltech professors. Brian gave an excellent description of his work, impressing the committee with the science and wowing them with his presentation.

At the end, Wal, wrapping up, said, "Well, today we have seen the debut of Kirshner, Junior."

Both Brian and I turned equal shades of crimson. I wonder who sat next to Brian at lunch. Perhaps some legendary figure from the past. I don't know because he must have been seated out at the edge of the room.

So, what time is it? On the face of it, a cosmic age of 14 billion years from a Hubble constant of 70 is in the same ballpark with the ages of the globular cluster stars at 12 billion years or the cooling time for the oldest white dwarfs stars in the Milky Way at 10 billion. If the universe has been expanding at a constant rate, then the cosmic ages seem concordant. Stars that formed a few billion years after the hot beginning are younger than the universe as a whole. This is good, because you should not be older than your mother.

But the concordance of ages is spoiled if the universe has been slowing down. In that case, the present expansion is a treacherous guide to the rate since the beginning of time. In fact, this was a very serious problem through the 1980s and 1990s. If the Hubble constant were really 80 kilometers per second per megaparsec, as

Figure 6.3. **Brian Schmidt explains the expanding photosphere method to his Ph.D. advisor in 1994.** The computer screen shows Schmidt's Hubble diagram for type II supernovae, derived using the expanding photosphere method to measure distances. Courtesy of Harvard News Office.

suggested by Wendy Freedman's team in their first report in 1994, and if Ω were equal to one, as many theorists believed, then deceleration would make the age of the universe embarrassingly small compared to the ages of stars. If the universe had been expanding more rapidly in the past, just like observing a tiring marathon runner, you would overestimate the time on the course if you ignored the slowing down. In this case, $\Omega = 1$ would imply a true cosmic age close to 8 billion years, which was not in good accord with evidence from stars. Something was wrong with this picture. Was it H_0 or was it Ω?

And finally, when do we get there? If Ω is low, we never do—the universe expands without limit. If Ω is one, universal expansion slows, but never turns around—we get close, but we never arrive. And if Ω is bigger than one, at some distant time in the future the universe will reach a maximum extent. We will have arrived, but we will also see the awful prospect of what lies ahead: universal contraction, the undoing of all the effects of hundreds of billions of years of cosmic change in a fiery Big Crunch. All of this sounds like the stuff of mythology, but we have slowly expanded the boundaries of measurement and rational discussion. The problem is not conceptual, it is quantitative: can we make the measurement well enough to trust the answer? And finally, those conclusions ignore the cosmological constant. If the total energy density of the world is made up of some matter that gravitates, and some dark energy that makes the universe spring apart, all bets are off.

7

a hot day in holmdel

Today we observe an expanding universe, with distances between galaxies stretching out according to Hubble's law. At the austere summits of remote mountains in Chile and Hawaii and Arizona, giant telescopes slowly gather photons from distant galaxies, building up the evidence for understanding an ancient and remote universe. But another important component of the universe was discovered in Holmdel, New Jersey, near Exit 114 off the Garden State Parkway. In this prosaic setting, Arno Penzias and Bob Wilson found that the universe is full of ancient light: glowing embers from the Big Bang.

More precisely, in 1965 they found a hiss of radio emission everywhere they pointed their radio antenna. We now know this emission has the spectrum an opaque object emits at a temperature of 2.725 ± 0.001 kelvins. That's 2.725 centigrade-sized degrees above absolute zero. Today, the universe is transparent, so photons can travel from distant galaxies to us without being absorbed. The light we see from galaxies is a complex mixture of emission from many different stars and gas clouds that carries subtle information about the composition and temperature and motion of the emitters. But the emission Penzias and Wilson discovered was much simpler—it comes almost exactly equally from all directions and its entire spectrum is described by just one number, the temperature. There are no details.

This gentle bath of low-energy photons is the relic of an earlier time when the universe was hot and opaque, so it behaved like an oven. When you heat an electric oven, the heating element emits infrared light, the cool walls absorb that light, and they warm up until they too begin to glow with infrared light. When an oven has been fully preheated, the thermostat switches off the heating element. Emission from the walls now fills the oven with an even glow of infrared light. When you put the dough in a baking pan and slide it into the oven, it absorbs energy from the infrared photons bouncing around inside the oven until it, too, approaches the temperature of the walls. Now you're cooking!

That's how you bake bread—dough warms up toward the temperature of the oven walls as it absorbs infrared light. Everything in an oven tends toward the same temperature. Bouncing photons guarantee that this equilibrium is enforced. The spectrum of photons inside the oven is determined only by the temperature, not by the chemical composition of the oven walls or the type of raisins in the dough. Ordinary kitchen ovens don't get hot enough for human eyes to see the walls glow, but a ceramic kiln or a well-kindled charcoal grill does. The red glow of coals in the heart of a fire is radiation of this type, and we all know that the color of the coals tells the temperature of the fire—dull red coals are cooler than bright orange ones. The cosmic microwave background is the glow from the hot Big Bang—but the temperature of 2.725 degrees above absolute zero means we don't detect this with our eyes: we need radio receivers like the one that Penzias and Wilson built.

Inside any region of the opaque universe, the same effect is at work—all the objects in an opaque universe come to the same temperature, because photons flying around at the speed of light ensure that any region that is a little cooler gets warmed up, while any region that was a little hotter gets cooled off. Penzias and Wilson detected photons that had their spectrum formed when the universe was opaque. Straightforward calculation shows the universe then had a temperature of at least 4000 kelvins.

So the cosmic microwave background photons observed in New Jersey come from a time when the universe was 1000 times hotter than it is today. These photons have stretched with the cosmic expansion by a factor of about 1000 since they last bounced off matter. Emitted as visible light, they have been degraded by expansion down to the low-energy photons that radio telescopes detect so well.

These photons fly through a transparent universe, carrying their image of the infant universe in all directions. When they were emitted, the scale of the universe was 1000 times smaller, the density of matter in the universe a billion times higher, and the temperature 1000 times hotter. Those photons show us what the universe was like when it was very young, just at the moment when it changed from being opaque, like the walls of an oven, to transparent, like a window.

This physical change in the universe at large from opaque to transparent results from the microscopic rearrangement of individual electrons and protons. When the universe was hot, the electrons and protons that make up ordinary matter were moving too fast to assemble into hydrogen atoms. Photons bouncing around had plenty of energy to rip apart any atom that did form. But, after about 300,000 years of expansion and cooling the warm post-Big Bang haze of matter and light finally cooled enough for electrons to give up their freedom. Electrons joined protons to form hydrogen atoms without being harassed by disruptive ultraviolet photons. Free electrons are good at scattering light; hydrogen atoms with electrons in bound orbits are much less effective: the hazy universe turned transparent when hydrogen atoms formed for the very first time.[1]

The cosmic microwave background (CMB) provides the most direct evidence that the universe had its origin in a hot Big Bang. This is not just an impression based on the expansion galaxies show, but a real physical change in the universe over time. The universe we see today has elaborated over cosmic time from a hot, opaque, evenly distributed soup into a cold, transparent, lumpy universe with galaxies, stars, planets, and people. The early universe was simple and predicable using straightforward physics. But once the universe turned transparent, things began to get interesting, complicated, and unpredictable. That's the messy realm of astronomy.

Detecting the cosmic microwave background was a major event for cosmology. A hot Big Bang had been contemplated by George Gamow and his students Hermann and Alpher decades earlier as a possible site for the synthesis of elements, but this never led to a search, and the site of manufacture for heavy elements was later identified in stars and supernovae. Even though Penzias and Wilson were not intending to find out anything about the universe, their measurement was so important that they received the Nobel Prize in Physics for 1978.[2]

But there is something curious about the uniformity of the CMB. The fuzzy horizon of the CMB is off in the distance in all directions 14 billion light-years away. And the temperature we see in any direction is 2.725 kelvins. But, spin on your heel, and you can also see 14 billion light-years in the opposite direction, where the temperature is also 2.725 kelvins.[3] Now, in an oven, things come to the same temperature because the photons from a warm region sap energy from the hot places and heat up the cool regions. But photons can only travel at the speed of light, and when they are bouncing around in a fog, they propagate even more slowly. The regions we see on opposite sides of the sky have *never* been able to exchange photons to even out differences. Why do they have the same temperature?

There's something odd about this. It's as if you traveled a billion light-years at 99.999% of the speed of light, landed on a planet, and found the inhabitants playing baseball. By exactly the rules of major-league baseball: no aluminum bats. It would make you wonder, if you were really our first emissary to this distant place, how they knew to play by the same rules as the Red Sox. So the question is, "How did the universe get so uniform?"

One idea that sounds wild and fanciful, but that is taken seriously by thoughtful people, is that the entire patch of the universe that we see in all directions was once small enough for photons to establish a single temperature. Then, due to an energy associated with empty space, the universe underwent a tremendous exponential expansion in which the scale of the universe increased by a factor of something like 10^{50} during the time around 10^{-35} seconds after the Big Bang. In this picture, during the "inflation era," the observable universe grew from a region so small that photons could cross it in the time available into something the size of a grapefruit. The precise numbers depend on the details of how particles and fields behave at energies that have never been observed by any particle accelerator on Earth, but the basic idea does not depend on these details. In this picture, today's expanding cosmic horizon is once again encountering regions that were once before in contact.

In other words, before inflation, the material in the observable universe was once in good thermal contact, like the interior of an oven. Then, during the inflationary era, the universe expanded exponentially, placing regions that were once in touch out of contact. Inflation ended somewhere around 10^{-35} seconds, then a lot of time (10^{17} seconds—a Hubble time!) passed. For each place, the observable patch of the universe grows—now we can see other parts of the universe billions of light-years away. When regions say hello again, 14 billion years later, they have the same temperature because they were in touch long, long ago, in the fraction of an instant before inflation got rolling.

This "inflation" idea sounds crazy. The fact that it is taken seriously by people who sit firmly in endowed chairs doesn't automatically make it right. But it has strong roots in the quantum world of particle physics and it does more than just resolve the "horizon problem" of a uniform temperature in parts of the universe that are just now getting in touch. Inflation makes this a neonatal ward reunion instead of a first-time meeting. Inflation also makes some firm predictions about departures from absolute smoothness and about the geometry of the universe. These predictions can be subjected to observational tests. If the predictions are not borne out, then the simplest version of the inflation idea can't be right.

If the predictions are confirmed, that doesn't necessarily mean inflation is the right picture. After all, there could be some other idea we haven't thought of yet that would also make these predictions. But if inflation keeps passing observational tests, it's not just sloppy logic to think we might be on the right track. It could have been shown wrong!

The physical mechanism for inflation has its roots in the weird world of quantum physics. One idea that has proved very fruitful in the quantum realm has been to think about the properties of empty space: the vacuum. In the subatomic realm, ordinary common sense ideas turn out to be worse than inadequate—they are just plain wrong. In the big world of things we can see with our own eyes, objects like a thrown baseball have a definite location at every instant, and motion that we can measure with a radar gun. But on the small scale of electrons and protons and below, these commonsense ideas of position and motion are replaced by a kind of intrinsic vagueness: the Heisenberg uncertainty principle says that you cannot know both the exact position and motion of something at the same time.

For big objects, this is not a practical issue, but on the subatomic scale, it is of the essence. The human scale is as big compared to the atomic scale as a star is compared to a human. You can't really expect to have a good feel for what things are like for an electron. We can't say an electron orbiting the proton in a hydrogen atom is exactly "there," with precisely such and such a motion, but are driven to more subtle formulations describing the probability of finding an electron in a given state.

For inflation, the weird idea is that the vacuum of empty space may have an energy associated with it. You may think empty space must have zero energy, but physics does not tell us that empty space must have zero energy. It's a little like looking at a topographic map of the Earth—the heights are given as the distance above sea level, but that leaves out the radius of the Earth. In the same way, physical events tell us about energy *differences*, but they don't tell us if there's an underlying floor of vacuum energy. There could be, either for a brief moment, or for a longer time, an energy of the vacuum that is not quite zero that lurks below all the measurements of energy differences that we make.

The effect of a vacuum energy in general relativity would be a "negative pressure" that makes the expansion of the universe accelerate. If the energy in the vacuum stays constant or just declines slowly enough, the rate of expansion is proportional to the size: it is literally an exponential growth, just like compound interest, and just like currency inflation. In December 1979, Alan Guth, a not-so-young postdoc in a temporary job (now Weisskopf Professor at MIT), was not thinking about career advancement as he rode his bicycle to work at the Stanford Linear Accelerator. He was thinking about what might happen if the universe got into a state where the vacuum energy wasn't zero. He was so eager to get to work that morning, to check out the consequences of his wild idea, that he set his personal best cycling time of 9 minutes, 32 seconds. After a few years of bruising price rises in the late 1970s, inflation was in the back of everybody's mind, even a mind as busy with other ideas as that of the other-worldly Guth. That's why this runaway expansion of the universe in the first 10^{-35} seconds is called inflationary cosmology.[4]

Physicists like this idea for the origin of the Big Bang. First, it comes from their turf: the world of theoretical particle physics, not the messy world of astronomical observation. "Scalar fields," like the field that produces inflation, are their bread and butter. Scalar fields give masses to the quarks that make up neutrons and protons. Particle physicists do not regard inventing such entities out of whole cloth as a strange way to think. They do this before breakfast. Second, it is mathematically elegant, and if truth is beauty, then beauty is truth and inflation must be the right model. Or, to put it more seriously, this is a powerful and attractive theoretical idea. Third, it accounts for known facts like the expanding universe and the uniform microwave background. But most important, it makes some predictions, at least in its most straightforward forms, that observers can test. Inflation spans the microscopic and the cosmic—it is audacious, esthetically appealing, and, best of all, we can find out if it is wrong.

One prediction of the simplest version of inflation is that the universe will have the geometry of flat space: that $\Omega = 1$. Even if the universe started out with some curvature, the tremendous expansion of the inflationary era would increase the radius of that curvature and force the geometry to become the geometry of flat space. If you take a region the size of a grapefruit and expand it to the size of the universe, the rind will be very, very flat. Or, as Guth says, "The value of omega will be driven to one with exquisite precision." So, if we can measure the effects of Ω, we can test whether this is true and find out if this version of inflation is wrong.[5]

A more subtle feature of inflation is that you can compute the character of variations in density from place to place in the universe. If quantum mechanics rules the first instants of the universe, then quantum uncertainty predicts there must be a range of values for the density of matter and energy that you measure when you sample different chunks of the universe. What this means is that the universe should contain a variety of density variations that resemble waves ranging from tiny little ripples to the longest waves that could fit into the cosmic horizon at every instant of the inflation era. These variations in energy density will leave an imprint on the cosmic microwave background that we can detect as subtle temperature differences from place to place in that smooth background, like a watermark on otherwise smooth bond paper.

These random variations would be the ultimate origin of the large density differences we see today in the distribution of galaxies. The action of gravity in the past 14 billion years amplifies those initial seeds into the jungle of cosmic ecology we observe today. We start from random fluctuations, gravity organizes matter to form galaxies and stars, nuclear physics elaborates the elements inside stars, and then the universe begins to get interesting, eventually making planets and people. So another test of inflation is to see whether people on a planet (Earth!) can see the predicted fluctuations in the microwave background.

Early measurements of the cosmic emission showed that the microwave background is smooth. Unlike the high-contrast galaxy distribution we see today, with dense clusters and yawning empty voids, at the time when the universe cooled and turned transparent, matter in the universe was almost exactly evenly distributed everywhere. Almost exactly, but not quite. Very careful measurement of the CMB from satellites, balloons, and ground-based instruments at very dry sites like the Atacama desert in Chile and the South Pole shows definite signatures of subtle variations in the brightness of the background.

The lumpiness in this cosmic soup is about one part in 100,000—that's like having a scoop that digs out $1000 in pennies from a tremendous penny jar, and getting the same answer every time, to the penny. That's really smooth. A baby's bottom is the colloquial standard of smoothness. Hands-on observations of my own children showed that a bottom has bumps of 0.1 millimeters on a span of 10 centimeters, so it's only smooth to one part in 1000—a human infant's skin is a hundred times rougher than the infant universe. And that's without diaper rash.

A map of these tiny variations reveals some important clues to the physical state of the universe when it was young. It shows the dense regions, destined to grow denser as gravity magnifies inequality and the low-density regions that are fated to lose out as time goes by. These tiny variations are the seeds that flower into the high-contrast bouquet of clusters and voids that we see today in galaxy surveys. Just as the rich get richer, the dense get denser through ruthless cosmic unfairness as gravity makes contrast grow.

The first map of these fluctuations in the early universe was made in 1992 by the Cosmic Background Explorer (COBE) satellite. Those early observations smeared together the measurements to an angular scale of about 6 degrees, about the angle your fist covers on the sky when you hold it at arm's length. Even in this blurry image of the sky, COBE definitely detected fluctuations of the general sort predicted by inflation. While this did not prove that the inflation model was right, it was a test that the model could have failed.[6]

We see an expanding universe, with the distance between galaxies stretching out over time. We see the relic glow of a time when the universe was young and smooth and hot. There is another piece of evidence that the universe we see today is the result of a hot Big Bang 14 billion years ago. That is the ubiquitous presence of helium, the second-simplest element, in stars of all ages. Helium is produced after inflation ends (if inflation really happens) in the hot, expanding universe.

There are degrees of audacity. Inflation is an extrapolation far beyond anything we're ever measured in a terrestrial laboratory. While it is an intriguing idea, it is a speculation. The inflation era corresponds to energies 10^{13} times larger than have been produced

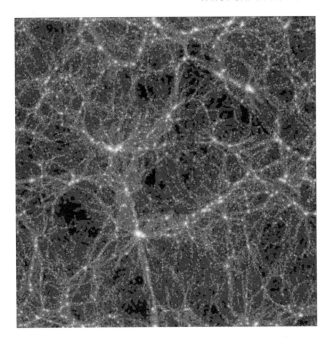

Figure 7.1. **The growth of structure.** Once baryons recombined, they could move under the force of gravity. Matter that could form galaxies, stars, planets, and people drained into the valleys that dark matter formed, as shown in these computer simulations. The distribution of luminous matter traces the presence of dark matter. Courtesy of The VIRGO Consortium.

in the most powerful particle accelerator on Earth. As we learn more about the subatomic world, as we continue to journey inward toward measuring the properties of the very small, inflation may or may not seem like such a great idea twenty years from now. But the world at 4000 kelvins or 40,000 kelvins or 40 million or even 40 billion kelvins is well within the scope of today's experimental physics.[7] We're not guessing about how electrons and protons and neutrons and neutrinos interact at these temperatures. This is the low-energy realm of nuclear reactions and, for good and for ill, we know how those reactions work in stars and in bombs. Thinking about a 100 billion degree opaque universe in the seconds after the Big Bang is not nearly such a big extrapolation as speculating what happened in the first 10^{-35} second! Our knowledge of the time when the universe was as hot as the inside of an exploding star is really quite secure.

Complex elements such as oxygen or iron are produced when stars generate energy or erupt in supernova explosions. We know from spectra that the oldest stars in our Milky Way galaxy have only 1/1000 as much of these elements as the sun does. This means that the abundance of these elements has been building up over time, like old shoes in the back of the closet. The exception to this is the element helium—the second element in the periodic table. Although helium is produced in stars as they fuse hydrogen, even the oldest stars have about as much helium as the sun. When we look at gas clouds in other galaxies, as Wal Sargent and Leonard Searle were doing in the early 1970s, following up lists of strange objects that Fritz Zwicky was compiling in his basement workroom at Caltech, they found that there are some galaxies with very little oxygen. Presumably these are the places where stars have done the least to enrich the mix, and these galaxies are closest to the composition that came out of the Big Bang itself. But even the most pristine gas cloud seems to have a dollop of about 25% of its mass in helium. This is a powerful clue that helium has not been building up over time in the same way as other elements. How did helium get a head start?

The answer to this riddle lies farther back in time than the era of stars, in the hot, dense Big Bang. The microwave background shows us an image of the universe when it was 1000 times cozier than it is today. If we dare to push back another factor of 1000 in cosmic scale, beyond the time we observe directly, the universe would have been a million times hotter than it is today. We can't see into that era, because the universe was opaque, but we do understand how things work at these temperatures and densities. We can't see into the center of the sun, but we know what's going on in there, and this is similar, if more remote. Going back another factor of a million beyond that stage is still within the realm of well-tested terrestrial physics. The universe would have been a nuclear furnace, fusing the lightest particles into helium. Or, more precisely, since the universe was on a one-way trip from hot to cold, a nuclear freezer, in which nuclei froze out once the temperature was low enough.[8]

At the end of the first few minutes after the Big Bang, as the temperature sank low enough for the simplest nuclei to stick together without being broken up by high-energy photons, there must have been a universal game of musical chairs. Every proton would have grabbed a neutron to form deuterium, and then in a few steps, the deuterium nuclei would form helium. A helium nucleus has two protons and two neutrons, so by computing the number of neutrons present when the universe was cool enough for deuterium to stick together, we can figure out how much helium would form in the expanding Big Bang. This works out to be about 25 percent of all the mass of ordinary matter. And that's just about what we see. When the numbers come out this close, the ideas have the ring of truth.

Even the first generation of stars would start out with a dowry of helium from the Big Bang. George Gamow started out with the aim of cooking the elements in the fireball of the Big Bang, but this source stumbles at the gap from helium to lithium that stars bridge by whacking three helium nuclei together to make carbon. We inherit carbon and oxygen and iron and gold from previous generations of stars, but helium is a legacy directly from the Big Bang itself.[9]

So we have good observational reasons to think the universe began about 14 billion years ago as a hot, dense Big Bang. After a brief early epoch of exponential expansion, the universe was a simple, hot, nearly uniform place. The element helium formed as that oven cooled. Before (re)combination, the growth of contrast, at least for ordinary matter like protons and electrons, was checked by the interactions of matter with light, which would act to smooth out any lumps. After recombination, hydrogen atoms made up most of the ordinary matter, and once the universe was transparent, gravity could begin to make ordinary matter grow lumpy. There must have been a first generation of stars in which nuclear reactions generated energy and made a start on the elements of the periodic table. Galaxies began to form out of the uneven distribution of matter, and big galaxies formed by gobbling up their little neighbors.

Our Milky Way would be the product of a long tree of mergers going back 13 billion years into the past. The sun and the Earth formed from the gas rich in iron and silicon and calcium and oxygen and carbon accumulated in our galaxy after 8 billion years of stellar burning. And here we are, living things made of carbon and calcium and iron, breathing oxygen, and looking back up the river of time toward our origins. This is a beautiful and simple picture of where we came from.

Of course, beautiful and simple are not always the same as "correct." If you look closely enough at the luminous fresco by Michelangelo that decorates the ceiling of the Sistine Chapel, you can begin to see the cracks, smudges, and gaps in the picture. In the same way, if you take a close look at this picture for the expanding universe, you can see places that need more work. This doesn't necessarily mean the framework is wrong, but it does mean we need to understand better what is in the universe and how the laws of physics play out to make the world around us.

One crack in the fresco is the measured amount of matter, and our curious inability to say precisely what the matter of the universe is made of. While inflation suggests that $\Omega = 1$, direct attempts to measure the matter of the universe indicate something different. Fritz Zwicky, irascible but prescient, showed in 1933 how to measure the mass associated with galaxies by measuring the speed of galaxies as they swarm in galaxy clusters. The more mass in a cluster, the faster the galaxies will move. Measure the motions of galaxies relative to the cluster redshift and infer the mass. This technique, and other effects that depend on mass that have been developed in recent decades, like the emission of X-rays from gas in clusters, or gravitational lensing in clusters, all point to the same result—the total mass that is clumped with galaxies is much larger than the mass of the stars emitting visible light, but much too small to give a gravitating mass density, Ω_m, equal to one. The best estimates give values of Ω_m closer to 0.3 ± 0.1.

A common approach to this problem of the contents of the universe, employed regularly over the last decade, but familiar since Biblical times, has been a heady mixture of skepticism mixed with flattery and a dash of pride. More than one theorist has said to me, with a thin-lipped smile, "Well, Bob, measuring the matter density and the expansion rate of the universe are very difficult things done by talented, but, let's be frank, fallible observational astronomers. Astronomers have been wrong before and may well be wrong now. Not all observations are correct. Since we know, from our highly developed esthetic sense, that Ω_m equal to one is the right answer, you observers should just go back and do the measurements again until you get it right."

We bring data down from the mountains on magnetic tape, not stone tablets, and there have been many false steps in building the observational picture of the universe. What has changed, but only in the last five years, is that the observations have become more certain, more telling, and the conclusions cannot be ignored even when the implications are quite uncomfortable. This has led to a surprising new synthesis of theory and observation, but only by inviting one of the old skeletons out of the closet: Λ the cosmological constant.

What makes the measurement of the matter content of the universe especially interesting is that even Ω_m of 0.3 demands that most of the matter in the universe is invisible and unfamiliar stuff. Put another way, $\Omega_m = 0.3 \pm 0.1$ is 7σ low compared to $\Omega_m = 1$, but big compared to the density you'd get by adding up the masses of all the visible stars that make galaxies shine. If you do that, you get only $\Omega = 0.005$. More generously, when you add in the mass of hot gas we see emitting X-rays and all the other matter we can detect directly, the sum is still only about one-tenth of the total mass we know is present in galaxy clusters. We know the mass is present because we see its gravitational effects, but we don't see light of any form being emitted or absorbed by this material. So we conclude that most of the matter in clusters, and presumably in the universe at large, is dark. Zwicky named this "*dunkle Materie*," dark matter. "Matter" because we know it is there. "Dark" because we can't see it. But having a name for something doesn't necessarily mean you know what it is. Or as Zwicky said in 1957, "It is not certain how these startling results must ultimately be interpreted."[10]

There is an even more curious problem with the nature of the dark matter, based on a combination of observation, reasonable physical theory, and current understanding of helium cooking in the Big Bang. That confluence of evidence suggests that most of the dark matter is *not* made of the neutrons and protons and electrons that make up our bodies, the Earth, and all the stars we see, but is mostly "matter" that is very different from the material world we know.

The argument is a bit subtle, but it leads to a very interesting conclusion. During the nuclear cooking that synthesizes helium in the first minutes of time, deuterium, the delicate isotope of hydrogen that has one neutron and one proton, plays a special role. Deuterium sets the moment when helium synthesis can begin. Helium gets assembled only after the universe cools enough that deuterium can survive the bath of gamma rays that is the cosmic background radiation in the early universe. Most of the deuterium nuclei then get locked up into helium nuclei, but a little is left over. The moment of helium synthesis passes as the universe expands and cools. Some stragglers of deuterium survive to become part of the gas in the universe we see today.

The amount of deuterium that survives the mad dash to assemble helium is small, but detectible. The leftover amount is very sensitive to the density of neutrons and protons in the universe at the time of helium assembly. So the amount of deuterium delivered by Big Bang cooking depends on Ω—the density. More precisely, it depends on Ω_b, the fraction of the universe that is made of baryons. "Baryon" comes from the Greek word for heavy—and this is apt since neutrons and protons are heavy compared to the leptons (from the Greek word for "light"), like the electron and the neutrino. Here's the curious fact: measurement of the amount of deuterium, seen in absorption lines formed in intergalactic gas clouds, shows that the amount of deuterium (several parts in 10^5) left over from the era of helium cooking is more than 10 times larger than you'd compute for $\Omega_b = 1$. The best estimate for Ω_b based on the residual deuterium is about 0.04 ± 0.01.

Quantities matter. If the amount of matter, Ω_m, is about 0.3, and the baryon density Ω_b is 0.04, 7 times smaller, then most of the matter in the universe cannot be baryons. Even if measuring errors and systematic errors have thrown both of these numbers off by a factor of two, we would still conclude that most of the dark matter in the universe cannot be anything made from neutrons and protons—the stuff of all the chemical elements, and of our own bodies. If we take this conclusion seriously, then we are *not* made of the kind of stuff that makes up most of the universe.

What's more, when we use our baryonic brains to try to think what most of the matter in the universe could be, there is one conspicuous candidate. We know of elusive particles that don't emit or absorb light and are not baryons: neutrinos. Neutrinos seem like a very good candidate for the dark matter, except for one thing. The problem with neutrinos as the gravitating matter that makes up most of the mass in the universe is that they have too little of precisely the one thing dark matter must have: mass. Lack of mass is a real drawback for something that is supposed to outweigh all the stars in the universe! There is evidence now from underground neutrino detectors that the mass of a neutrino is not quite zero, so neutrinos do make a small contribution to the total Ω_m of about 0.003. A neater universe crafted by Occam's razor might have just one form of dark matter, but our extravagant universe apparently must have at least three: some dark baryons, a pinch of neutrino mass, but mostly something else. Instead of a minimalist universe, we seem to live in a rococo one: we have everything you can think of, and more than you can think of. Perhaps we should not be so quick to use Occam's razor to reject wild ideas: we need even wilder ones to interpret these startling results.

If we follow this chain of argument, most of the universe is in a form of dark matter that isn't baryons and isn't neutrinos. We know what it isn't but we don't know what it is. Theoretical particle physics has produced some possible candidates with whimsical names like the axion and the neutralino. These particles may have the right properties to be the dark matter, but at present they have the distinct disadvantage that they have not yet been discovered! Particle physicists are rightfully proud of the role that powerful theoretical ideas have played in predicting the existence of particles that have later been found (like Dirac's prophecy of the positron—the antimatter clone of the electron). But it doesn't seem unreasonable to wait for terrestrial experiments to show that these particles actually exist and have the right mass before asserting too confidently that they make up most of the universe.

If the dark matter is something like a neutrino, only with a lot more mass, those particles would be everywhere. Since they don't interact by the strong force that glues nuclei and they don't interact by the electrical force that makes it hard for people to walk through walls, these "weakly interacting massive particles" (WIMPs to the wags who dub these things) would be present in the room where you read this book. As the Earth orbits the sun, the sun orbits the center of the Milky Way, and M31 tugs the Milky Way in its direction we would be drifting through a mist of WIMPs just as we are drifting through the photons of the cosmic microwave background. You can detect the microwave background from anywhere, and you could find the dark matter just by catching one of these particles as it drifts through your laboratory.

Now, just as the academic prestige of the inflation theorists doesn't prove they are right, the fact that people have built experiments to detect WIMPs doesn't prove that most of the mass in the universe is in this weird form. But it does show that competent people take these arguments seriously enough to test the ideas by observation. As a scientist, you really have control of only one resource: your own time. When professors and postdocs and graduate students spend years to build a delicate WIMP-catching apparatus, and set it up, not at a beautiful mountaintop in Chile, or even off the Garden State Parkway in New Jersey, but deep in an oppressive abandoned iron mine in the middle of nowhere Minnesota, you know they are serious about trying to find out what the world is made of.

Cosmic timescales pose the most difficult problem for a universe with Ω_m equal to one. Gravitation slows cosmic expansion, but the amount depends on Ω_m. In the low-Ω_m case, you can correctly compute the cosmic age from the present rate of expansion, $t_0 = 1/H_0$. Recall our mythical marathoner Eddie, who computed the time elapsed in the Boston Marathon without a watch. He measured distance and velocity for various runners *assuming all of them ran at a steady speed* from the start in Hopkinton to the finish line on

Boylston Street. This is just like a low-density universe, where gravitation doesn't slow the expansion.

If the universe does have an appreciable mass density, the relation between the present rate of expansion and the actual elapsed time since the Big Bang is a little less simple. Gravitation slows expansion, making the Hubble time an overestimate for the age of the universe. Estimating the age of the universe from the local expansion rate, the Hubble constant we measure in the local patch out to 1 or 2 billion light-years, is equivalent to looking only at the last miles of the Boston Marathon. You don't know what the runners were doing earlier, so you just assume that the present is like the past, and make your best estimate. But it ain't necessarily so. If the runners are actually slowing down, but you watch them only over the last mile, you will overestimate how long they've been running the course. If some poor footsore devil limps the last mile in 10 minutes, you might think they've been out on the course for 26 miles × 10 minutes/mile = 260 minutes = 4 hours and 20 minutes. But maybe they were churning along fine at 7 minutes per mile until they hit the wall on Heartbreak Hill and they've been slowing down ever since. Observing those aching survivors only at the end of their travail will lead you to overestimate the actual time they've been suffering out on the course.

Similarly, if mass has been decelerating the cosmos, then the universe, like somebody who turned gray at 35, is younger than you think from a first glance (always check the eyebrows!). If you start out with Ω_m closer to one, the slowing-down effect gets larger. The boundary of this ever-slowing expansion is $\Omega_m = 1.000000. \ldots$ In that case, when you compute the effect of expansion and deceleration, the age of the universe turns out to be exactly two-thirds of the age you would infer from the present rate of expansion. The real elapsed time since the Big Bang is just two-thirds of the Hubble time. In symbols, we could write, $t_o = 2/3 \ (1/H_o)$.

If gravitation has been slowing cosmic expansion, the real age of the universe would be *younger* than 14 billion years. Deceleration would reduce this to 9 billion years—significantly shorter than the 12 billion years estimated for ages of the oldest globular clusters or white dwarfs. This would be embarrassing. Even taking into account the uncertainty in the ages of the oldest stars of 1 billion years, this would be a 3σ discrepancy. Gauss says that only happens by chance one time in 370, so, if the numbers are right, there's a 99.7 percent chance that there's a real problem with the cosmic ages. Globular clusters should not be older than the universe in which they reside! Common sense suggests that this much deceleration can't be present, even though $\Omega_m = 1$ apparently demands it. This is definitely a crack in the fresco! Or, to put it in a more positive light, what we know about the ages of stars helps separate the one real universe we actually live in from the many that are mathematically possible.

Appeals to common sense are not good enough. We should look for effects to measure from direct observation, not esthetics, or even logic, whether the universe has or has not been decelerating. The best way to do this is to use powerful telescopes to look deep into the past to see how cosmic expansion has changed over time. In recent years, we have used supernovae, detected halfway back to the Big Bang, to trace the history of cosmic expansion and measure its change.

A value of $\Omega_m = 1$ is the razor's edge. If Ω_m is even slightly more than one, say 1.001, then the expansion will eventually stop, reverse, and become a contraction. If the universe started out in a Big Bang, then a universe with Ω_m greater than one will eventually end up in a gnaB giB, back in that unimaginably hot and dense state. All the elaboration of the universe would be reversed—stars would evaporate back into gas, nuclei eventually melt back into the simple particles out of which they are made, and the wonderful complexity of the world would be erased. It's not a pretty thought, but we shouldn't expect the universe to care what we think.

Although the cosmological constant was exiled to a theoretical leper colony after the 1930s, it is worth exploring how Λ affects cosmic ages. Einstein invented Λ to balance out gravitation to produce a static, eternal universe. Eternal is an age. Infinitely old. De Sitter noticed that Λ would make a massless universe accelerate, and Eddington suspected that Slipher's observations of the recession velocities for spiral nebulae were the effect of Λ at work, perhaps accelerating the galaxies from rest.

But there are more possibilities. If you have some dark matter, Ω_m, and some dark energy, Ω_Λ, mathematical solutions to Einstein's equations have complicated and interesting properties. If Ω_m and Ω_Λ have just the right values, the universe would expand, slow down under the influence of gravitating matter to almost zero speed, and the universe could loiter there before the repulsive effect of Λ would initiate an era of accelerating expansion. Before Hubble established the velocity–distance relation, this model had the feature, then thought to be desirable, of a long static period with no expansion, as Einstein had imposed in 1917.

The essential point is that a universe with both Ω and Λ has a more complicated relation between the present rate of expansion, H_o, and the cosmic age, t_o. In the decelerating phase, the universe would be younger than $1/H_o$. In the quasi-static phase, H_o would be near zero, and the universe would appear, like ill-mannered party guests who have overstayed their welcome, as if it would linger there forever, even though it had a finite age. In the accelerating phase, the then-current rate of expansion H_o would be above the average, like a runner sprinting for the finish, and the elapsed time since the Big Bang could be longer than you'd compute from $1/H_o$. Like a game show host who has had a facelift, an accelerating universe would appear younger than it really is.

When Eddington was talking about loopholes to reconcile the (wildly mistaken) long ages of stars in the 1930s with the (wildly mistaken) short expansion age of the universe in 1931, he was thinking of the way that adjusting Λ could fix this problem. In polite circles, and even in astronomical discussions, using Λ to reconcile problems with timescales went the way of spats on shoes. They were kept in the attic as a relic of the 1920s, brought out on special occasions just for fun, but never worn at a serious event—until about 1996 when some fashion leaders tried them on at Princeton. We may all be wearing spats again.

A gravitating mass density equal to one has attractive mathematical properties, just as Ω_Λ has been regarded as ugly. Following Einstein's example, theorists look for the simplest formulations, with confidence that nature will follow (or, more precisely, precede) their good taste. If the mathematics looks beautiful, theorists take that as a sign they are on the right track. A "standard cold dark matter" universe with Ω_m = 1 has a greater esthetic appeal than a low-density universe in which the density just keeps getting lower, so that Ω_m drifts toward zero. And it has a better look than a high-density universe (Ω_m greater than one) in which the density eventually grows uncontrollably when the universe begins to contract. Ugh! But, like the porridge, chair, and bed that Goldilocks prefers, the Ω_m = 1 universe is just right, and a universe with Ω_m = 1 *stays* a universe with Ω_m = 1 even as the universe expands and slows. For Ω_m of exactly one, the density decreases at just the right rate so that the ratio of the actual density to the critical density remains constant. In the inflation picture, there's an inescapable reason for Ω to be one: the immense expansion drives Ω inexorably to this value by ironing out any curvature.

This esthetic argument grips the theoretical mind like a bear hug and has been very close to the center of the cosmological discussion for the past 20 years. Particle physicists call their picture of the realm of quarks and the forces that bind them "the standard model." Looking for a little reflected glory, theoretical cosmologists have referred to the Ω_m = 1 possibility as the "standard cold dark matter model." This was a good rhetorical device. But, as we shall see, it has two problems. One is the cosmic timescale. If the universe has been decelerating in the way a universe dominated by gravitating matter requires, then the age of the universe comes into conflict with the ages of stars. The other is that measured masses of galaxies give Ω_m, the density of dark matter associated with galaxies, well below one. So if the total Ω really is one, but the density of gravitating matter Ω_m is not one, something else must contribute very significantly to the density of the universe. What could that be?

One possibility is something that gravitates, but does not cluster with the galaxies. If matter is distributed smoothly, it could elude our measurements in clusters. This would be "hot dark matter" where the individual particles have such high speeds they don't fall in to the deep troughs of galaxy clusters. The problem with hot dark matter is that, if it is important, it would smear out the growing structures of the universe too much to make the lumpy universe we see in redshift surveys. Elaborate numerical calculations of the way that structure grows in the universe show that hot dark matter would make a much smoother universe than the one we observe. The large-scale distribution of galaxies seen in big redshift surveys simply can't be matched if hot dark matter is the most important constituent. In an extravagant universe, where all the possibilities seem to be present, we can't rule out *some* dark matter of this type, but we have good evidence written in the sky that there is not enough to make $\Omega = 1$.

Another possibility is that it could be the cosmological constant. The mass equivalent of the dark energy contributes to the total Ω as Ω_Λ. It could help make the universe flat, but would not show up in measurements of the matter density Ω_m. You could have a dollop of dark matter and a dollop of dark energy to make a total Ω of one. But there are good reasons to be wary of this siren's call. Are you sure you want to use something Einstein grew to regret?

Only in the last few years, as observations have grown more telling, have we been able to move from a debate based on esthetics to a discussion based on evidence. Observations have dragged us reluctantly toward accepting the view that the universe is dominated by the strange properties of empty space. After all, Einstein did say of the cosmological constant, "observations will enable us in the future . . . to determine its value." The future is now.

8

learning to swim

Our small brains decode the messages encrypted in ancient light to build an orderly picture for the universe that matches the observations and obeys the local laws of physics. A correct scientific idea had better agree with the physical and astronomical facts as we know them. But because our current knowledge is incomplete, it's not smart to impose too strict a censorship on ideas. Common sense isn't always the best guide because the real universe is more bizarre than anyone dares to imagine. On the other hand, ideas are not useful just because they are wild. They must match the facts. The cosmological constant has always been a wild idea. As invented by Einstein in 1917, it was used to account for a static universe. In the 1930s, this wild idea was discarded by most astronomers because it was not needed to match the observed fact of an expanding universe. But now we have a broader concept of what Λ might be: we think of it as a dark vacuum energy with negative pressure. After 70 years of excluding Λ, new facts not only permit, but require something like the cosmological constant.

What are those facts? Since the 1930s, cosmic expansion has been a fact, though obtaining precise and accurate measurements of the present rate of cosmic expansion has provided astronomers with decades of difficult and contentious work. Since 1965, the cosmic microwave background radiation's remnant glow has been a firm fact that any physical picture for the early days of the universe must match. The evidence glimpsed in the 1930s from the furious zooming about of galaxies in clusters shows that galaxies have much more mass than meets the eye: galaxies are trapped in invisible pits of dark matter. The evidence from helium and from deuterium, the heavy isotope of hydrogen, has become another fact that any picture must match. Deuterium measurements place such a low ceiling on the density of baryons that confidence in the picture of the freeze-out from a hot Big Bang has led to the strange view that most of the matter in the universe is not anything we know from the periodic table of the elements and is definitely not the stuff we are made of.

Then we have some astronomical facts. These are often inferences based on a long chain of measurement and reasoning. Because these facts result from such a complex set of observations and ideas, it is hard to know exactly what measuring errors and systematic errors lurk behind the digits. The way to find out is to make measurements by a variety of methods—if they disagree, you can stage a debate so proponents can make arguments about which method has the biggest errors, but when they agree, then you may be getting close to the truth. With those cautions, it is reasonable to say we know that the oldest stars in our own galaxy are about 12 ± 1 billion years old. Also, we observe a value for the present rate of cosmic expansion, the Hubble constant, of about 70 ± 7 kilometers per second per megaparsec. And, when we measure the mass associated with clusters of galaxies, we find the cosmic density, expressed as a fraction of the critical density, gives $\Omega_m = 0.3 \pm 0.1$. These facts give the background for observing cosmic acceleration, which we observe directly by measuring the apparent brightness of supernovae at large redshifts.

Einstein's theory of gravity, applied to the universe as a whole, lets us predict what we will see when we look deeply into the past. For the past 50 years, astronomers have been trying to test these predictions to find out what kind of universe we live in. Telescopes observe the distant past. An important observational test is to measure the change over cosmic time in the rate of cosmic expansion.

The wild card in this confrontation of theory with evidence is that the predictions are simple only if there is no cosmological constant. For the past fifty years, almost every discussion of these cosmological tests starts with a brief disclaimer—that the results apply for $\Lambda = 0$. Given the universal distaste for blundering, those less talented than Einstein have stayed well to windward of the cosmological constant. Only the convergence of powerful facts could convince a skeptical community that Λ really is necessary.

The 200-inch Hale Telescope at Palomar Mountain was put into action in the early 1950s. During the decades while it was the world's most powerful telescope, until it was supplanted by the 10-meter (400-inch) Keck telescope in 1993, hundreds of nights were assigned to the problem of determining the deceleration of the universe from observations. In 1961, Allan Sandage spelled out how these tests could be done. Despite heroic efforts, this observational program using the brightness of galaxies to map the history of cosmic expansion stalled—nobody produced credible evidence for changes over time in the expansion rate. But the seeds of success were sown. As astronomers slowly developed a base of knowledge about supernova explosions, we created the tools and techniques for measuring the acceleration of the universe.

In the past five years, as a result of improved instruments like the Keck and the Hubble Space Telescope, diligent accumulation of data on nearby supernovae, and a concerted effort by two international teams to measure supernovae halfway across the universe, we are beginning to paint a new, messy, and wild picture for the cosmos. It's an extravagant universe. To match all the evidence, we need a universe that has ordinary matter, glowing and dark; dark matter of at least three kinds: baryons, neutrinos, and weakly interacting massive particles (WIMPs); and a large dollop of dark energy whose negative pressure drove the inflation era and another, much longer-lived dark energy that drives cosmic acceleration now. You would be unwise to believe such a baroque mixture, which seems to violate common sense, Occam's razor, and the boundaries of good taste, except that there are lines of evidence, from direct measurement of the matter density, from the concordance of cosmic ages, and from the subtle watermark of manufacture observed in the background radiation, all of which converge on a view that the universe now has a preponderance of dark energy. Dark energy, which might be the cosmological constant, or something that changes with time, has moved from being a wild idea, not really fit for serious discussion, to an essential feature of our present view of the universe. How did this happen?

The first step in developing the evidence for the accelerating universe has been to develop a reliable ruler for measuring distances in the universe. Today's best tool is the explosion of a white dwarf star as a type Ia supernova. In the 1940s, Walter Baade, working at the Mount Wilson Observatory in Pasadena, began to compile measurements of supernova brightnesses. He and Fritz Zwicky had worked together to establish that supernovae were a genuine phenomenon, different from ordinary novae, in which the stupendous energy release signaled the death of a star. Zwicky used cobbled-up cameras, and then, after 1936, his new 18-inch Schmidt wide-field telescope at Palomar Mountain to find supernovae. Baade and another Mount Wilson astronomer, Rudolph Minkowski, took spectra of the supernovae at Mount Wilson. Their goal was to find out from empirical observation what supernovae were and then to puzzle out from those clues what their physical origin might be.

Since a supernova is as powerful as a few billion suns, Baade recognized that supernovae might be useful in measuring extragalactic distances. Just as Hubble had used cepheid variables to chart the distances to nearby galaxies, Baade reasoned that the supernovae might be useful yardsticks to intermediate distances, large enough to provide an independent calibration of the Hubble constant.

When Baade looked into this question in 1938, he found that supernovae were not super good as standard candles. The typical scatter in brightness was a sigma of about a factor of three. In 1938, supernovae were a coarse ruler for measuring cosmic distances. But today, they are the very best "standard candle" for cosmology. What changed?

First, Minkowski made a very important contribution, which has been elaborated in the past 60 years. He looked at the spectra of supernovae, which convey information about the chemical composition and the expansion speed of the stellar debris. The spectra of the original handful of supernovae were very strange compared to

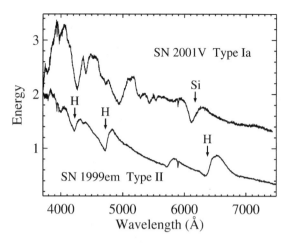

Figure 8.1. **Spectra of type I and type II supernovae**. Type I supernovae do not have lines of hydrogen while Type II supernovae have prominent hydrogen lines. Although this does not exhaust the possibilities, with type Ib (and type Ic) being introduced later, most supernova spectra we observe are of these two general types. Courtesy of Tom Matheson, Harvard-Smithsonian Center for Astrophysics.

any ordinary star, but similar from one event to another. As Minkowski put it in 1939, "the spectra of all supernovae are practically identical."[1] But in 1940, Minkowski found a supernova that broke the mold: "the spectrum of this supernova is entirely different from that of any nova or supernova previously observed."[2] SN 1940B had strong and easily identified lines of hydrogen. It was a supernova of a different kind. Minkowski's observation split supernovae into two classes, type I and type II.

Type I was the original type with the mysterious, but uniform, spectrum. The prototype was SN 1937C, an especially bright object in a nearby galaxy for which Minkowski obtained spectra out to 339 days after maximum light. Even if you didn't understand the origin of the spectrum, if it was the same as SN 1937C, then it was a type I supernova. Type II was the type with hydrogen lines. By sorting out the supernova types, Minkowski put us on the track to understanding that there is more than one way to explode a star, and to using spectra to sharpen up supernova samples. If you toss out the type II supernovae, the ones that remain are more similar to one another, and better standard candles. It's a little like trying to determine the average height of sixth-grade boys. You do a much better job if you make sure there aren't any girls mixed in because the girls are much taller at that age!

Now an empirical method is a good thing, and Minkowski's description of the spectra was clear enough for others to identify supernova spectra in the same way. What was not clear, at first, was the physical origin of supernova explosions. An empirical method that you don't understand is not as good as one that has a foundation.

By the 1960s, using the principles of nuclear physics, Willy Fowler and Fred Hoyle shed some light on the origin of type I and type II supernovae. They traced the history of nuclear burning in stars of different masses. Low-mass stars, up to about 8 times the mass of the sun, end up as white dwarfs, with a core that is made of carbon and oxygen, or in the case of the most massive progenitors, oxygen, neon, and magnesium. The white dwarf does not ignite because the star is held up by quantum mechanical effects. White dwarf stars are potential thermonuclear bombs because they have unburned fuel, but, like a stick of dynamite, they are harmless unless detonated. Hoyle and Fowler identified the type I supernovae as the nuclear explosion of a white dwarf, an event that might be precipitated by added mass from a binary companion. This provided a theoretical underpinning to uniformity—since there is a fixed upper mass limit to white dwarfs of 1.4 solar masses as worked out by Chandrasekhar, a uniform energy output might come from explosions of identical stars at that maximum mass.

The life history of more massive stars is different because they fuse carbon and oxygen without detonating. They burn oxygen into sulfur and silicon and, eventually, fuse all the way to iron. Then, at the nadir of nuclear binding, they collapse. Hoyle and Fowler weren't too clear on the details, but they surmised that these collapsing events inside a star with hydrogen on the outside, with an immense release of gravitational energy, could produce the type II supernovae that Minkowski had identified.

In 1970, as a skinny, red-headed kid of 21, I arrived as a graduate student at Caltech. I was assigned to Bev Oke, one of the astronomy faculty, for a research job to supplement my National Science Foundation fellowship. Oke, a friendly, modest, red-headed Canadian, had applied advances in light detectors to the job of measuring spectra at the 200-inch Hale telescope. When I showed up in his office on the second floor of Robinson Lab, he asked blandly, "Well, what do you want to do?"

I didn't really know, but I knew enough to avoid three areas of astronomy that I thought were really dull. One was the measurement of parallax, which demands more patience and precision than I possess. Another was studying dust, which is messy stuff, whose properties are exceptionally hard to measure and interpret. And the third was spectral classification, which has an empirical quality of making fine distinctions that resembles philately. Curiously, the study of supernovae has drawn me into each of these areas and each of them has been essential to building up the picture of the accelerating universe.

As a senior at Harvard, I had enjoyed working on ultraviolet emission from the sun with Bob Noyes. Bob had been a graduate student at Caltech a decade earlier and he encouraged me to go to Pasadena. He wrote a letter of recommendation. I don't know if he was honestly ignorant of my slightly erratic academic record, or whether he explored the outer bounds of puffery, but the letter worked and, much to my surprise, I was admitted to the astronomy program at Caltech. Hal Zirin, a Caltech professor who studied the sun (known as Captain Corona to the students), much later told me that he had lobbied hard for a graduate student who might work on the sun. Me. Well, in the end, I didn't, but it worked out all right. Bev Oke was giving me a chance to use data from the world's largest telescope. I just wanted to avoid parallax, dust, and spectral classification.

Remembering the fun I'd had working on the Crab Nebula, a supernova remnant, I said, "I'd be interested in working on supernovae." The legend was that Caltech professors had so much telescope time they would take data and then put it away like fine wine until the moment for its analysis ripened. In a quintessential Caltech moment, Oke opened a drawer in his desk, and pulled out a fistful of Kodak yellow cardboard jackets containing spectra of supernovae. "Here," he said, "see what you can do with these." I had no idea what to do with them, but I wasn't going to admit that on the first day of graduate school.

In that bundle there were spectra of type I supernovae and spectra of type II supernovae recorded on photographic plates. Oke had also invented a new instrument, the multichannel spectrophotometer (the "multichannel"), which made simultaneous quantitative measurements of the light from an object at 32 different wavelengths. This was a big step up from earlier instruments that could make a similar measurement at just one wavelength, though a long step from today's instruments that make 1000 such measurements of 100 objects in a single observation. With the world's best telescope and the world's best instrument, we would have to be dull indeed not to do something useful.

The first bunch of data Oke handed me included a set of observations of SN 1970G, a type II supernova in the nearby galaxy M101. He also had several observations of type II supernovae with the multichannel. The 200-inch users, including Chip Arp, Maarten Schmidt, Leonard Searle, Wal Sargent, and Jim Gunn, following the good examples of Baade and Minkowski, cooperated to get good coverage of the changing spectra of supernovae during the weeks when an object was bright. In fact, their motivation was a little stronger than altruism—there was a sense of *noblesse oblige*. Supernovae had been understood first in Pasadena, studied best in Pasadena, and it was natural for people in Pasadena with the world's best instrument to contribute to this topic, which seemed important in its own right, if not yet useful to cosmology. And now I was holding a fistful of supernova spectra. I had an obligation to understand them, even if I had no idea how to proceed.

I took all this grist down to my office in Robinson Lab. As a beginning student, I was placed in the second sub-basement of the building, where all the offices had numbers that began with 00, amusing the James Bond fans, though the only thing you were likely to kill was yourself. With work. To get to 0013, I had to go past another of the sub-basement suites, where there was a very strange and forbidding old man, wearing an eye patch as he worked away on a plate-measuring machine. He looked like a pirate. It was Fritz Zwicky.

Zwicky, the astrophysical swashbuckler who named the supernovae and the dark matter, charted the galaxy clusters, and launched the first interplanetary ball bearing. Zwicky, who claimed his "Morphological Method" was the greatest contribution to human thought since Pascal. Zwicky, at age 72, a terrifying spectacle for a fledgling graduate student who maybe ought to be studying the sun instead of Zwicky's own subject, supernovae. Fritz was wearing an eye patch to help look through the single eyepiece of a measuring machine where he was grinding away compiling his great catalog of galaxies and clusters of galaxies. He was tall and gaunt. His speech was as intimidating as his looks.

At that time, my wife was a substitute teacher. She would get calls before 6 A.M., telling her to become Miss Jones, third grade teacher at the Burbank School by 7:15. Awakened by these academic alarms, I would get up and walk over to Robinson Lab. Arriving before 7 A.M. is unusual in any academic setting, but at an astronomy department, the night owls usually showed up around noon and worked until midnight (I guess—how would I know?). But no matter how early I arrived, Zwicky was already there.

He began to talk to me briefly each day. He usually launched into bitter vituperation in a spicy Swiss–German accent, aimed at the current staff, including my advisor, Bev Oke.

"Those spherical bastards threw me off the 200 goddam-inch telescope!" he fumed. "Made up a special rule. No observing after the age of 70! Grrrr, them I could crush!"

A spherical bastard was "a bastard any way you looked at it." Or sometimes the injustice was more widespread.

"In 1933, I told those no-good spherical bastards that supernovae make the neutron stars. Now they find these damn pulsars and nobody gives me the credit."

Or "Quasars? Quasars? Maarten Schmidt and his goddamn quasars. They are objects Hades, by the Morphological Method predicted!!"

Figure 8.2. **Fritz Zwicky in 1971.** Here Fritz demonstrates the symmetry of a spherical bastard, "A bastard any way you look at it." Photo by Floyd Clark, courtesy of the Archives, California Institute of Technology.

These set-piece speeches blasting the Caltech faculty were shocking, subversive, and wickedly amusing at first. There was a large but finite number of them. They became familiar, then tedious, then a little embarrassing. Zwicky used these packaged diatribes as "questions" after a colloquium talk on any topic. So after a talk on the magnetic fields of white dwarfs, or galaxy dynamics revealing the dark matter, or the chemical composition of extragalactic gas clouds, we would once again learn of the injustice of quasar nomenclature, eliciting inward (and sometimes outward) groans in the audience.

Sometimes Zwicky would give me advice:

"Always get here before the Americans" (advice I could not possibly adhere to!).

Sometimes he would pose conundrums:

"Do you know how to get the 200-inch to give diffraction-limited images?"

I had to admit that I did not. As I understood it, the telescope's imaging was limited by the blurring effects of temperature inhomogeneities in the Earth's atmosphere. The atmospheric limit was about 50 times worse than the theoretical limit given by the mirror's size and the wavelength of light. It seemed like a sensible answer to me, suitable for the Ph.D. oral exam I was preparing to take.

"Hah!" Zwicky's face contorted with scorn. "Hah! You're just like the rest of those low flying shit-eaters! No, No, No! You fly a jet over the dome at the speed of sound! Then you use the shock wave like a knife edge. Those bastards never let me do it!"

I nodded, having only the vaguest idea what this enraged man was shouting about, but hoping to get to my office for a few hours of quiet work. I had to catch up with the night owls.

One morning Fritz seemed to be in orbit.

"Never mind the Bolsheviks and their so-called Sputnik. I, Fritz Zwicky, launched the first interplanetary probe!"

I was too amazed to inquire further. But one day, years later, I had an hour to kill in Alamagordo, New Mexico. The choices are limited. I recommend the New Mexico Museum of Space History. Upstairs beyond the gift shop selling inedible "astronaut ice cream," on the wall of the International Space Hall of Fame, there was a bronze plaque of Fritz Zwicky. Just like one in Cooperstown of Ted Williams. Fritz had been telling the truth! An Aerobee rocket launched at White Sands, NM on the night of 15 October 1957 carried a shaped explosive charge in its nose. After ascending 53 miles in 91 seconds, the explosive was detonated, blasting out luminous pellets at more than 9 miles per second, fast enough not just to orbit Earth, but to travel indefinitely out into the solar system. Fritz was not making this up.[3]

Even though Zwicky had written the book on supernova classification, I never told him I was working on supernovae—it seemed too dangerous. And he was too wrapped up in his own sense of injustice to bother asking. I don't think he ever asked my name.

But with Fritz in the next room, I felt some weight of history leaning on me. I learned to sort the SN I from the SN II. Like everybody else in the previous 40 years, I couldn't identify most of the absorption lines in the type I spectra, so I put those aside. The type II supernova spectra were more promising because even a beginner could understand what was going on. It was hydrogen, after all, that made the type II spectrum. I used the hydrogen lines to try to understand how the mass was distributed in the atmosphere of the exploding star. This might give a clue to the state of the star when it blew up. That seemed worth doing.

I was making some progress on understanding the atmospheres of type II supernovae when Bev Oke was invited to a winter workshop on supernovae at the Kitt Peak National Observatory in Tucson in February 1972. He suggested to the organizers that they should invite me, too. In his own quiet way, Bev was a very good advisor. His sharp sense of smell for a good scientific opportunity always put a student in a position to succeed, but Bev would rarely tell you what to do next. Sink? Swim? That part was up to you. But he'd take you to the beach.

I was delighted to go to Tucson where many of the hotshots in the field would be present. It was a great chance for a rookie to meet the All-Stars. Jerry Ostriker, the brilliant Princeton theorist, was there, full of new ideas about neutron stars, and Stirling Colgate, the wild-man physicist from Los Alamos who knew how to blow things up, and Craig Wheeler, already one of the best at connecting supernovae with the stars that make them. Our host, Leo Goldberg, had been the Harvard College Observatory Director when I was there, and was now the Director of Kitt Peak, no longer deploring paucity, but allocating plenty, and, completing his liberation from the Harvard faculty, no longer wearing a tie! Rudolph Minkowski was there, a living legend from Mount Wilson days, a pioneer of supernova studies, looking a little like a gray walrus with a brushy moustache, sagely puffing on a pipe.

Goldberg presided over a conference dinner in downtown Tucson. For entertainment, some guy from Livermore did magic tricks with a piece of rope, cutting it, but revealing it to be whole. Scientists, like everybody else only more so, don't believe in magic; we believe in evidence and reason, so the conflict between the evidence of our eyes and our faith in reason made us admire his deception twice as much. Or maybe it was the wine.

As the party broke up, I joined Craig Wheeler and Jerry Ostriker to walk the mile or so back to the University of Arizona campus. As we were approaching the campus, near the geometrically significant address of Euclid and University, a group of students cruising by in a 1965 Mustang found three astronomers oddly provocative. Perhaps it was Jerry's enthusiastic reply to their jeers. I think he said, "Free Angela Davis." Anyway, they stopped the car, carefully put down their six-packs of Lucky Lager, and rambled over to confront us. Craig had the collar of his Oxford-cloth shirt ripped, and Jerry had his wire-rimmed glasses broken again ("my optometrist will be cross with me") while I was wrestling with a pretty strong guy. He probably did not know I had been the 137-pound runner-up in the Harvard freshman intramurals, but I didn't feel compelled to inform him that my body was a deadly weapon. I was ahead on points, and executed a neat take-down, but something felt funny when my shoulder hit the sidewalk. I learned in an instant that cement is stiffer than a wrestling mat. Then I hit him in the fist with my lip, and they all fled, fearing dry cleaning bills.

The next morning, with my arm in a sling, I talked about the atmospheres of type II supernovae on the sunwashed patio at the Kitt Peak offices. There's something about a separated shoulder that takes the zip out of a presentation. Maybe it was the painkiller, or perhaps the inability to gesture vigorously. I started out by reviewing the data on hydrogen lines in type II supernovae. I showed how the data we had for SN 1970G indicated that as time passed, the velocity decreased. This did not mean the gas was slowing down—it meant we were seeing deeper into the star, where the velocities were lower. It was a way to reconstruct the mass distribution on the outside of the exploding star. Minkowski, 77 years old, was sitting in the front row, puffing on his pipe. He quickly grew impatient with this introductory material, put down his pipe, and growled in heavily German-accented English, "Ve know all dis." It was not a good start.

A more useful suggestion came back in Pasadena from Leonard Searle, one of the staff astronomers at the Carnegie Observatories (and later its director.) Genial Leonard had cooperated in getting data for SN 1970G, and he noticed that the multichannel data from the supernova photosphere (the surface where light escapes) defined a beautiful continuum—just like the blackbody spectrum from any opaque object. Wouldn't it be possible, Leonard asked, to use the information from the hydrogen lines, which gave the velocity, together with the multichannel scans, which could give a temperature, to work out the size of the supernova photosphere at several times and compute the distance to M101? Leonard's suggestion was to use supernova data alone to find the distance to the galaxy in which the supernova had exploded. I worked this out, using the data Oke and others had gathered at Palomar. Though Leonard Searle was right in principle, the problem was a little more complicated than it seemed at first. Another Caltech graduate student, John Kwan (now an astronomy professor at the University of Massachusetts), contributed ideas and worked out theoretical issues where I got stuck. We computed distances to M101 and NGC 1058 that were completely independent of all the intermediate steps in the extragalactic distance scale. Since the redshifts for those galaxies were well known, and part of the overall cosmic expansion, we felt justified in computing the ratio of velocity to distance, the Hubble constant. For this work, we found a value of the Hubble constant, H_o, of 60 ± 15 kilometers per second per megaparsec.[4]

At the same time, Allan Sandage, up on Santa Barbara Street, and Sandage's Swiss colleague from Basel, Gustav Tammann, had been working on distances to the very same galaxies, M101 and NGC 1058, using empirical methods that calibrated the properties of galaxies. Those two galaxies were too distant for 1970s technology to detect the individual cepheids. Sandage and Tammann were embroiled in a vigorous debate about the Hubble constant with Gerard de Vaucouleurs of the University of Texas. In the 1970s, de Vaucouleurs maintained that the evidence favored a high value of

H_0, around 80 or 90, while Sandage and Tammann stoutly maintained that 55 was the right answer. Each group claimed a precision that ruled out the answer given by the other. John Kwan and I had stepped into an arena already soaked with bad blood by heavyweight gladiators. At first, they were glad to see us. Tammann sent me a nice note, congratulating us on getting the right answer.

While the universe doesn't care what we think, we do. And Allan Sandage thought that our distances based on the expanding photospheres of type II supernovae were close enough to his to be a pretty good result and evidence against the misguided Parisian in a ten-gallon hat. So he regarded us as possible allies in resisting the falsehoods being issued from Austin. My own view was less dogmatic—I had no stake in the outcome, we were just trying to measure a number and that's the one we got. In the long run, I had confidence we'd find out what was going on. Then we could move on to error and confusion on a new set of questions.

Sandage's view seemed much more emotional—perhaps as Hubble's only student, and the world's leading practitioner of practical cosmology, he felt responsible for the Hubble constant and the Hubble time coming out right and making sense. Much later, in 1994, Ron Eastman, Brian Schmidt, and I used a larger set of data and the expanding photosphere method (EPM) to find $H_0 = 73 \pm 8$ kilometers per second per megaparsec, which was 2 sigma away from 55. Not so close. Sandage took a personal view of the Hubble constant—if you disagreed with him, you must be wrong, and possibly malicious. And if you changed from agreement to disagreement, you must be treacherous or stupid or both. At that time, I was the department chair at Harvard, and we invited Sandage to come to Cambridge to give a talk about his work on the Hubble constant. Sandage wrote back, declining. He said his mother had taught him not to talk to the village idiot.

The expanding photosphere method was a parallax— number 1 on my list of things not to do, and it also led to a confrontation with interstellar dust, number 2 on my list of things to avoid. Dust between the stars has been a bugaboo for astronomy for a century. Correct understanding of the size and shape of the Milky Way was hindered for decades until people worked out the effects of obscuring matter. One important clue to the presence of dust is that it absorbs blue light more effectively than red light. The signature of interstellar dust is "reddening." This resembles the effect you see at sunset, where the setting sun looks dimmer and much redder than the noonday sun, because the light traverses a longer path through the atmosphere and the atmosphere scatters and absorbs the sun's blue light, making the sun look red. When an astronomer sees a familiar type of object, but its color is unusually red, the first thought is that dust is responsible. Could dust be a problem for the EPM distances? (Figure 4.1 shows dimming in the direction of the center of the Milky Way.)

Dust doesn't make much difference to the distances derived from the expanding photospheres. Dust in our galaxy or in the galaxy where the supernova (formerly) resided absorbs light. This makes the supernova appear dimmer, so, other things being equal, you would mistakenly assign it a larger distance than the real one. However, since the dust removes more blue light than red light, it also makes the supernova appear redder. If the supernova's light is reddened, you would mistakenly assign the supernova a cooler temperature than it actually has, since cooler objects emit redder light. In the arithmetic of the EPM, this red color makes you think the supernova is closer than it really is. The two effects very nearly balance, so that the error you make because the supernova is dim is corrected by the error you make because of the change in color. By good fortune, dust doesn't create a big systematic error for type II supernova distances found through the expanding photosphere method. But the lesson was to think carefully about dust, or you might make a systematic error so large (and avoidable) someone else might call it a mistake.

In May of 1972, Charlie Kowal was using the 18-inch Schmidt at Palomar to search for supernovae. Zwicky's old telescope was fine for this work, and Charlie made a regular patrol of nearby galaxies where the wide field of the little Schmidt made it the best tool for the job. Tipping the telescope as far to the south as was prudent, Charlie exposed a film at the Centuarus group of galaxies, centered on NGC 5236, a big fat spiral galaxy with evidence of star formation and a history of producing supernovae. At the same time, he got an image of the insignificant little neighboring galaxy NGC 5253 for free.

When he developed that film, he placed it on top of an older film on a light box, aligning the two so that every dot that was in both epochs appeared double. Scanning the film by eye, one dot jumped out at him from the thousands on the film. It was a plump, solo dot—present in one film, but not the other. Separating the films, he saw it was tonight's film with the new object. Charlie had discovered another supernova. It was his job to discover supernovae, but that didn't make it less fun. And this was a really good one.

This was supernova 1972E in the galaxy NGC 5253. It was the brightest supernova in 35 years, since SN1937C, the one studied so well by Minkowski at Mount Wilson. Discovered at Palomar, SN 1972E was studied thoroughly at Palomar, using the multichannel scanner on the 200-inch, where it took only a few minutes to get a fabulous spectrum. What's more, there was a new telescope at Palomar, a 60-inch telescope that had been more-or-less finished, but not yet scheduled for observations. Since the multichannel was not going to be mounted on the 200-inch every night in May, Bev Oke thought it would be a good idea for somebody to go up to Palomar for a couple of weeks to make observations of SN 1972E on the new 60-inch. Even though this was a single-channel scanner, 32 times slower, on the 60-inch, with 10 times less collecting area, it would be good to get data every night. Was there a graduate student interested in supernovae and looking for a thesis project who wanted to do this? I raised my hand.

Bev Oke drove me up the mountain in his gray MGB hatchback. He was a careful driver, but he enjoyed the curves up Palomar Mountain more than I did. When we got to the observatory, one of the technicians saw two redheaded guys whose age differed by about 20 years getting out of Oke's car.

"Is that your son?" the electroniker asked Oke.

"Nope," Oke explained.

SN 1972E was a type Ia supernova, very similar to SN 1937C studied so carefully by Minkowski 35 years earlier. But now we had a beautiful set of modern digital data that covered the whole range from the ultraviolet to the near infrared. The 60-inch observations

I was making were 300 times slower than the observations Oke obtained at the 200-inch Big Eye. One minute of observation at the 200-inch collected as much information as 5 hours at the 60-inch. But in May 1972, SN 1972E was bright enough that I could get good data in a few hours. The 200-inch was overkill.

From Palomar, this supernova in the constellation Centaurus was scraping the southern horizon. At southern latitudes, in Chile, Pat Osmer was also observing SN 1972E. Pat had finished his Ph.D. at Caltech a few years earlier and was on the staff at Cerro Tololo Inter-American Observatory (CTIO). Pat was making observations very similar to mine with the Cerro Tololo 60-inch telescope at that excellent site. Even though SN 1972E was far in the south for us, our Palomar data set was more complete than Pat's for May when the supernova was bright, and, as the supernova faded in June and July, the speed advantage of the 200-inch made a huge difference. We compiled the best-ever record of the complex and mysterious spectrum of a type I supernova. That summer, Pat dropped by to show us his supernova spectra. They were good. Then we unfurled our massive set of observations. They were very good. Pat grew quiet and a little glum. This is the way they liked it in the old days at Caltech—the Big Eye blew the competition away.

When Subramanyan Chandrasekhar visited Caltech, he courteously took an hour to go to lunch at the Athenaeum, Caltech's Faculty Club, with the graduate students. A cerebral, slender man, Chandrasekhar was a formidable figure in theoretical astrophysics, whose career at Cambridge started with debates with Eddington and became a legend at the University of Chicago.

"Why," he politely asked the assembled group of six graduate students as they enjoyed their free lunch, "have you chosen to study at Caltech?" When nobody responded for about 30 milliseconds, I spoke up.

"Oh," I said, "that's easy. Caltech has the 200-inch."

He looked at me skeptically. "Really? You chose to come here because of a machine? How odd. I should have thought the faculty would matter."

In 1973, the International Astronomical Union had its once-every-three-years meeting in Sydney, Australia. Bev Oke was invited to give a review talk about supernovae to the whole General Assembly, since everybody wanted to see what we'd been doing with SN 1972E. He didn't want to go, but he suggested that I would be a good substitute. This was another good chance to meet the pros, only this time from all over the world, and from all fields. By good fortune and being at the right place at the right time (the right distance from NGC 5253 for the light, which had been traveling for 12 million years, to arrive at Earth in May 1972, just as I was looking for a thesis project), I was standing on the stage in front of 1500 astronomers, pretending to be an authority on supernovae.

But the most interesting aspect of studying SN 1972E came later. More than a year after the explosion, the supernova had faded so much that only the 200-inch could take spectra of it. Jim Gunn, then a young professor at Caltech, and Bev Oke integrated for hours to get the last observations. As a supernova expands, it eventually turns transparent, and you can see in toward the center of the explosion. Type I supernovae rise to a maximum brightness and then fade, quickly in the first month, then more slowly. After about two months, the brightness of the supernova tracks the radioactive decay rate of ^{56}Co, the isotope of cobalt that has 27 protons and 29 neutrons.

The theoretical idea is that a SN I is the thermonuclear detonation of a white dwarf. This implies that elements near iron in the periodic table (like cobalt) are produced in the explosion. An exploding white dwarf should blast out about 0.6 solar masses of elements near iron. When you consider that present-day gas in our galaxy has one iron atom for every 10^4 atoms of hydrogen, the sudden addition of 10^{55} iron nuclei from a type Ia supernova is a very important source of iron for the galaxy.

This is a good story, but we'd like to check the details of the prediction against the observations to see if it is true (or, more precisely, to see if it is false). Nuclear physics theory predicts that, in the conditions that prevail deep inside an exploding white dwarf, the most likely iron-peak product is ^{56}Ni, the isotope of nickel that has 28 protons and 28 neutrons. This is radioactive, with a half-life of 6 days. As nickel decays, it emits energy that helps make the supernova glow. When we catch a supernova rising toward maximum light, which takes about 20 days, or fading in the month after the peak, most of the energy comes from this radioactive decay. The decay product of ^{56}Ni is ^{56}Co, which has a 77-day half-life. So the long, slow decline in brightness that characterizes SN I is, in theory, due to the subsequent decay of cobalt into stable iron. Is this right?

Observations at early times showed that iron was accumulating at the expense of cobalt, and observations at late times also help test the idea. First of all, the spectra that Gunn and Oke obtained with the multichannel showed that the light curve continued to decline as predicted for at least 700 days. This made it plausible that the energy released from cobalt turning into iron was responsible. Even more telling was the spectrum, which showed four broad peaks. What were they? If we were seeing iron from the core of the explosion, heated by radioactive decay, then it seemed plausible that the spectrum at late times ought to be made up of emission lines of iron.

I had just received an HP-45 calculator for my twenty-fourth birthday, so I merrily sat down to compute what the spectrum of iron would look like under these conditions. I did a pretty crude job, but sometimes good enough is good enough. After one afternoon, it was clear that when you added up the emission from all the lines of iron atoms that were missing one electron, there was a good match with three of the four bumps in the late-time spectrum of SN 1972E. Bev Oke suggested looking at the contribution from iron that was missing two electrons, but I couldn't find a good compilation of the atomic data, so I wrote up the paper with the results we already had. I should have listened to my advisor. Tim Axelrod, a graduate student at Santa Cruz working with supernova wizard Stan Woosley, did the calculation right, including other forms of iron, and showed that the feature I could not account for was indeed due to iron stripped of two electrons.

I also should have included Jim Gunn on the list of authors for this paper—he had contributed many hours of his precious time on the 200-inch to these heroic observations. When I finally woke up to this gaffe a few years later, I sheepishly said to Jim that we should have made him an author of the late-time spectrum paper. Though he never said anything at the time, or in the intervening years, he hadn't forgotten. Jim smiled a bit and said, "Yes, Robert, you should have." Two lessons learned: listen to your advisor, and give credit where it is due. Observers have a choice of how to use their time and it is reasonable for people to be included in the published results even if the data itself is their only contribution. The most recent paper one of my graduate students wrote on type I supernova light curves had (because he listened to me!) 42 authors, each of whom had contributed some data.

The spectrum of SN 1972E was very similar to SN 1937C, the classic type I prototype that Minkowski had observed. The new data helped build the legend that all type I supernovae are the same. If their spectra were the same, and they all came from white dwarfs of the same mass, perhaps it would be a good idea to revisit their use as standard candles for measuring distances in the universe. Charlie Kowal, the Palomar supernova searcher, had compiled the data in 1968. His result was better than Baade's, but not by much. Supernovae of type I bounced around the inverse-square line with a scatter of around 70 percent, so assuming that SN I were all the same would lead to errors of about 35 percent in the distance of each host galaxy. This was better than Baade had found, and mildly encouraging for measuring cosmic expansion, but not good enough for measuring cosmic deceleration.

At the same time, Sandage, and independently Oke and Gunn, were trying to perfect the use of giant elliptical galaxies as distance indicators, to push the Hubble diagram out to distances that would reveal cosmic deceleration. In the early 1970s that seemed like a more promising path than using supernovae. The giant elliptical galaxies were brighter than supernovae by a factor of 30, and though they were extended, fuzzy objects, Gunn had devised an exceptionally clever way to deal with those complexities in the data that he and Oke were gathering with the multichannel.

A few years later, this massive effort to measure cosmic deceleration from the Hubble diagram for giant elliptical galaxies began to lose traction. The problem wasn't the measurements, difficult as they were, the problem was the galaxies. Galaxies are made of stars, and stars change over time. If there were high-mass, fast-evolving stars in a galaxy, you might expect the galaxy to be somewhat brighter when it was young. As time goes by, the massive stars would become supernovae, then wink out. On the other hand, galaxies are collections of stars that seem to form in groups and clusters. Though the stars don't collide, the galaxies can interact, and even swallow one another. Galactic cannibalism was probably most important for the big bright elliptical galaxies that people were using as standard candles. If a galaxy had grown over time, then it would have been dimmer in the past. Which was more important, stellar evolution that made galaxies brighter in the past or cannibalism that made them dimmer? Nobody knew, and the uncertainty in the properties of the galaxies was larger than the expected effects due to cosmic deceleration. By the 1980s, it was clear that another path needed to be found to crack this problem. Some people turned to supernovae.

9

getting it first

In the 1970s, supernovae were on the list of possible tools for cosmology, but not at the top of the list. In 1977, Bob Wagoner, a theoretical astrophysicist at Stanford, extended the idea of using expanding photospheres that John Kwan and I had timidly applied to nearby galaxies (where we actually had data) to cosmological distances (where there were no data!). This is OK for a theorist—it helps to illuminate the path we should be taking, not just pave the one we are on. Wagoner asked whether you could detect the effects of cosmic deceleration by applying the same method to type II supernovae at large redshifts. He showed that, in principle, you could, because deceleration would affect the relation between redshift and distance.

Hubble's law, with the redshift proportional to distance, is only an approximation to the whole story of cosmic expansion. It is almost exactly true nearby, but not necessarily true over a large fraction of the observable universe. Wagoner showed you could learn about the cosmology by making good observations of very distant supernovae. The difficulties were purely technical—in 1977 our telescopes and instruments were not up to the task of gathering the data you need at distances where cosmology makes a difference. Even though we know how useful these measurements would be, and the Keck 10-meter and HST are much more powerful than the 200-inch, there aren't yet any distances to SN II derived from expanding photospheres that bear on the question of cosmic deceleration. But there will be. Someday.

So attention has focused on using the SN Ia as distance indicators. The mythology, based on a few good examples, was that all SN I were identical. There were a few exceptions to this general rule—individual objects that didn't fit the pattern. None of these "peculiar" SN I was as well observed as the prototypes SN 1937C and SN 1972E, so it was hard to know whether the unusual features were genuine, or perhaps artifacts of marginal data. For example, I observed SN 1975A, which looked like a garden-variety SN I except it was missing the absorption dip at a wavelength of 6150 angstroms right where there was a very strong line in SN 1972E and SN 1937C. David Branch, at the University of Oklahoma, had made some headway in identifying the lines in SN I spectra, and he knew the missing line was due to the element silicon. Was this important, or just an insignificant variation on a well-established theme? There was a handful of similar cases sprinkled in among the accumulated data on SN I spectra and light curves. Did this detail in the spectrum matter? I had entered the twilight world of spectral classification, number 3 on my list of subjects to avoid.

This puzzle began to be solved in 1985. Alex Filippenko, who had been a graduate student with Wal Sargent at Caltech, was then a Miller Fellow at Berkeley. He and Wal were at Palomar taking spectra of galaxies that have strange emission lines. These emissions may well signal the presence of a massive black hole in the galaxy's center. New instruments at the 200-inch made it possible to get quantitative digital data not just at one tiny spot on the sky, as with the multichannel, but at 100 locations lined up along a narrow rectangular slit. Usually, you line up the slit so that the object you want to study is in the center, then the rest of the slit provides 99 excellent measurements of the spectrum of the sky. This is important for measuring very faint objects, because the night sky is bright, and the objects of interest are sometimes only 1 percent as bright as the sky. You need to do an exceptionally precise job of subtracting 100 units of sky light from 101 units of sky plus object to measure your target. Another way to use a long slit is to rotate it so you catch two objects at once—you get a spectrum of each, without using any additional telescope time. This makes you feel clever and virtuous.

On the night of 27 February 1985 when Alex slewed the Big Eye to NGC 4618, he noticed something odd. The picture he had brought to the telescope, his finding chart, showed the bright star-like nucleus of the galaxy, but there was a second starlike blob visible in the TV image from the telescope. Being curious, and hoping to be clever and virtuous as well, Alex carefully rotated the instrument so the entrance slit covered both the galaxy nucleus he had come to study and the new star that wasn't on his finding chart. The spectrum showed that the new object was a supernova of an unprecedented kind. SN 1985F showed huge powerful emission lines of oxygen and calcium. Filippenko and Sargent made a plausible case that this was the late-time stage of the explosion of a massive star. Oxygen and calcium in the star's midsection don't collapse into the neutron star, but are blasted out in the supernova explosion. But it was definitely not a regulation type II supernova because it had no hydrogen. It was something new.

Before long, a number of us, including Filippenko and my postdocs Alan Uomoto and Eric Schlegel, working together with Craig Wheeler's team in Texas, began to see what was going on. The two mysteries—peculiar type I spectra with missing silicon lines, and the weird spectrum of SN 1985F, were really two aspects of just one new thing. If you observed a peculiar SN I long enough, as it turned transparent in a few months, its spectrum changed into something like SN 1985F. A plausible story for these objects was that they were massive stars, as Filippenko and Sargent had inferred, that exploded after they shed their hydrogen-rich envelopes. They would be massive stars with a core collapse, but without the big coating of unburned hydrogen that makes SN II so distinctive and easy to understand.

In the tradition of stellar classification, we give these things names. They definitely were not type II because they have no hydrogen in their spectra near maximum light. But they are not like SN 1937C or SN 1972E where the late-time spectrum is dominated by iron emission from an incinerated white dwarf. To keep things straight, we decided to call the original type I supernovae type Ia, and this new class type Ib.

These names made sense to us working in the field, but like many astronomical names, they drive physicists crazy.[1] Physicists want to know why the objects that are similar inside, and that operate by the same physical mechanism—gravitational core collapse—are called by different names, type II and type Ib, while the objects that are fundamentally different, one having a thermonuclear explosion in a white dwarf and the other a core collapse in a star without a hydrogen atmosphere, are called by similar names, type Ia and type Ib. The short answer is, that's what we call them. And the reason is, the classification is based on the appearance of the spectrum at maximum light, which Minkowski measured in 1940, long before the details of the physical mechanisms were understood. And the spectrum will still be the spectrum, even if our understanding of the mechanisms changes. I was not only deeply engaged in stellar classification—on my list of things never to do—I was defending it!

Sorting out the SN Ia from the SN Ib had some unexpected benefits. In our own Milky Way galaxy, we see the remnants of supernovae that have gone off in the past 20,000 years, and there is a small handful of more recent ones where there is a written record of the explosion. Tycho's supernova of 1572, first observed in a pre-dinner walk and confirmed by country people going past in carriages, fits neatly with the physical picture for a SN Ia. But another young remnant, Cassiopeia A, was a puzzle. Observations showed that Cassiopeia A was expanding rapidly. If you extrapolated backward to ask when these star shreds began flying outward, the answer was around A.D. 1670. So, if there was a seventeenth-century supernova in our own galaxy in the constellation Cassiopeia, easily visible to all Europe, why hadn't anybody seen it?

When I was a postdoc at Kitt Peak, Roger Chevalier (now a professor at the University of Virginia) and I used the new 4-meter telescope to take spectra of the fast-moving gas in Cassiopeia A. The spectrum showed powerful emission lines of oxygen in some cases and oxygen plus calcium, argon, and sulfur in others, but no hydrogen. This looked to us like the innards of a massive star, perhaps 15 times the mass of the sun, which would have layers where helium had fused to make carbon and oxygen, and then further layers where the fusion of oxygen made calcium, argon, and sulfur. But where was the hydrogen that makes up most of any 15 solar mass star?

The observations of SN Ib suggested that some massive stars lose their hydrogen envelopes in a wind before they explode. Was Cassiopeia A the remnant of a type Ib supernova? Rob Fesen, once my graduate student at Michigan and now a professor at Dartmouth, has been pursuing this question. He has found some fast-moving hydrogen—probably the last bit on the surface of the pre-supernova star at the time when the blast wave roared through. A supernova without its hydrogen overcoat would be intrinsically dim, and if hidden behind some dust, might not be so easily observed. It might do nothing in the nighttime. Maybe Cassiopeia A came from a type Ib supernova explosion.

The idea that SN Ib come from massive stars that have lost most of their hydrogen envelopes has proved very helpful—this explains the peculiar SN I spectra seen at maximum light, accounts for the appearance of SN 1985F's spectrum at late times, and connects Cassiopeia A, a 300-year-old event in the Milky Way, with its extragalactic siblings. When you account for three phenomena with one idea, that's a good sign. The SN Ib story is important for cosmic acceleration because once you sift out the SN Ib, what remains as SN Ia is more homogeneous. These objects had been sneaking on to the lists of SN I, but now the masquerade was over.

By the late 1980s, Sandage and Tammann were working hard to make SN I tools for cosmology and were gearing up to measuring the Hubble constant with them. The idea was to use the Hubble Space Telescope to observe cepheids in the nearest galaxies that were the sites of type Ia supernovae. Tammann's student in Basel, Bruno Leibundgut, was compiling all the reliable supernova light curves and building up a template composed of observations of many SN I. Tammann and I had a good conversation about this in a beer garden near Munich at a meeting about SN 1987A. Beautifully dressed amid the carelessly casual scientists, and chain-smoking

cigarettes with a long, black cigarette holder, Tammann was a conspicuous and very effective ambassador for the program he and
Sandage were carrying out. When you differed with Sandage, he
conveyed a sense of betrayal, but with Tammann, it was a vigorous
debate without the personal overtones. Besides, we were both
thinking about buying Saabs.

"You also are interested in Saabs? I dream of Saabs."

But I got a little excited describing our data on SN Ib, insisting
that they were something new. Tammann countered that the homogeneity of the SN I was absolutely established both empirically and
theoretically and there must be some mistake that was leading to
this horrible and absolutely anti-Copernican view. What with the
Weissbier and the jet lag, somehow an emphatic gesture of mine
propelled a gigantic *mass* of Bavarian beer onto Gustav Tammann's
beautiful white suit. I knew then that the time for discussion had
passed and it was time to go to bed!

Tammann was depending on the homogeneity of SN I, and at
first he resisted the idea of a new subclass. He wrote to me, ob-
jecting that these so-called SN Ib were fainter and redder, so the
simplest explanation a competent astronomer would think of was
that they were heavily obscured by dust. I mildly pointed out that
the new classification was based on spectra and that the emission
lines at late times in the new type were very different from the SN
I lines we all knew and loved: oxygen and calcium instead of iron.
So the SN Ib were not SN Ia dimmed and reddened by dust, but
objects of different chemistry.

In the end, the introduction of this new class was a good thing
for the homogeneity of SN Ia and for the program of Tammann and
Sandage. Using Leibundgut's template to stitch together data from
various objects, the scatter from the inverse square line was still
about 40 percent, or about 20 percent in the distance. This was
very useful for measuring the Hubble constant, even from a small
number of objects, but still not good enough for measuring cosmic
deceleration with SN Ia.

Here's how we expected the deceleration measurement to
work. Suppose you wanted to distinguish between a universe with
$\Omega_m = 0$ and a universe with $\Omega_m = 1$. For the moment, assume, like

every respectable astronomer since 1930, that the cosmological constant is zero, so properties of the universe are determined only by gravitating matter: $\Omega = \Omega_m$. The $\Omega = 1$ universe has precisely the required mass density to slow the expansion at all times, but not enough to make the expansion stop and reverse. How would that deceleration show up in a measurement you can make with supernovae?

Nearby, the apparent brightness of a supernova drops off as the inverse square of the distance. And nearby the redshift is proportional to the distance. So when we plot the brightness against the redshift, we find, if the objects are good standard candles, that the points scatter around a line.

That's just geometry, without deceleration. But we can also compute the effect of changes in the cosmic expansion rate during the time the light is en route from its origin in an exploding supernova to its detection at a telescope. For nearby supernovae, this effect can be ignored, but for very distant ones, it can reveal the history of cosmic expansion. In a high density universe ($\Omega = 1$), while the light is on its way from the supernova explosion to you, the universe is expanding, but the expansion is slowing down. To get from the explosion to your telescope, the photon travels a smaller distance than in the empty universe ($\Omega = 0$) case where the universe is coasting. As you might guess, since the light travels a smaller distance, the supernova appears brighter. Of course, you have to do this computation correctly, taking into account the curvature of space, the stretching of time, and the shifting of photon energy, as well as deceleration, but the basic idea is right—if the universe is slowing down, then a distant supernova will appear brighter than if the universe is expanding at a constant rate.

Perhaps an analogy will help make this vivid. A bright, energetic kid with reddish hair I once knew used to throw snowballs at schoolbuses. This antisocial behavior may have been the result of a certain generalized boredom with third grade or perhaps of aspirations toward a major league career with the Red Sox. In any case, the impact of throwing snowballs at a receding bus, like sending photons across an expanding universe, depends on whether the bus is traveling at a constant rate, or slowing down for the stop sign up ahead. If you throw at a bus that is cruising along, the snowball takes longer to get there and makes a less satisfying splat. If you throw at a bus that is slowing down, the snowball travels a smaller distance and makes a resounding thud. Once the thud was so loud, the bus began to move backward as the irate driver sought the culprit. But I digress.

So the sign of a decelerating universe, as expected for $\Omega = 1$ (and $\Lambda = 0$), is that distant supernovae will appear a little brighter than they would in a universe where $\Omega = 0$ (and $\Lambda = 0$). For completeness, we should think about what happens when Λ is not zero. If Λ is not zero, then the recent history of the universe could include a period of acceleration. Looking at a supernova in an accelerating universe means looking at photons that travel an extra distance, so the supernova would look fainter at the same redshift. This is something like hurling a snowball at a bus that is accelerating away after it drops you off. If the snowball catches up at all, it barely sticks.

Supernovae at the same redshift in an $\Omega = 1$ universe should look brighter than in an $\Omega = 0$ universe. But how much brighter? If supernovae are not such great standard candles, there is a big natural variation from one to the next at any redshift. Then a small effect of cosmology will be masked by big differences in exploding white dwarfs. In general, as you go to bigger redshifts, the cosmological effects become more important (though for Λ different from zero, this can be a little intricate), but at the same time, the measuring errors for very faint objects become large. So the quantitative question is whether the SN Ia make good enough standard candles to reveal the effects of cosmology at redshifts where you can actually make the measurements. The difference in apparent brightness between an $\Omega = 1$ universe and an $\Omega = 0$ universe amounts to about 25% in apparent brightness for identical objects at a redshift of 0.5. At this redshift, the detectors of the late 1980s on 2-meter telescopes were adequate for making the brightness measurements, and the spectrographs on 4- and 5-meter telescopes had a sporting chance of getting a spectrum that could tell you the redshift. Maybe it was time to start searching for high-redshift supernovae.

But if the variation from one SN Ia to the next is 40 percent, then you need to observe many objects to get a well-determined average value. Generally speaking, Gauss tells you that the uncertainty in the average is reduced by the square root of the number of objects. So, if you decided you wanted a 3σ result to tell the difference between $\Omega = 1$ and $\Omega = 0$, you'd need at least 25 distant supernovae. That's because you want the final error to be about $25\%/3 = 8\%$. But if each supernova has a 40 percent error, to beat the errors down to 8 percent by sheer numbers of objects, you would need $(40/8)^2 = 5^2 = 25$ supernovae. This is the brute-force method.

But if you could reduce the scatter by understanding the supernovae better, you could make a meaningful measurement with fewer objects. Since the number of supernovae needed goes like the *square* of the measuring error, you save a lot of effort by improving that measuring error. If you make the errors half as large, you can get an equally valid result with only one-fourth the number of supernovae. This seemed to me like the smart place to put our effort.

People were definitely thinking about how to do cosmology with supernovae. You could measure the history of cosmic expansion by observing the relation between brightness and redshift. In 1979, 11 years before the Hubble Space Telescope reached orbit, Stirling Colgate wrote a paper in *The Astrophysical Journal* sketching a way to use the HST both to find and measure supernovae. Reading this paper today, you find that some paragraphs are wrong and many are unrealistic, but taken as a whole, the paper makes a good case for trying to do this problem when the technology ripens. That same year, Gustav Tammann gave a more careful analysis of what might be done to measure cosmological numbers with the Hubble Space Telescope by using supernovae.[2]

If supernovae were all alike, they would line up perfectly along the inverse square line of brightness and redshift in a Hubble diagram. But some supernovae are intrinsically brighter than others, so even at the same redshift, they won't lie exactly on this line. The observed scatter around the inverse square line, which measures how much supernovae vary from one to the next, was about 40 percent in brightness. SN Ia became a Rorschach test. Optimists like

Colgate saw a reasonable chance of getting some information about cosmology from supernovae. Optimists hoped that people would find a way to mold the supernovae into better standard candles, or hoped you could average the observations of many supernovae to extract a meaningful signal from noisy data. But in any case, by plunging in, you would find out what unanticipated problems stood in your way and you could begin to solve them.

Others thought that chasing high-redshift supernovae was a sink of time until you had a way to shrink the errors and make SN Ia much better standard candles. The pessimists (or "realists" as we prefer to be called) thought that the best place to focus effort was at low redshifts where understanding supernovae better might help make them more effective for cosmology, just as cleaning up the SN Ib problem had helped. Even if it didn't help with cosmology, studying supernovae would lead to understanding interesting astronomical events that are important in the origin of the elements and in the formation of galaxies. Generally speaking, the optimists were theorists or newcomers who had not worked long in the supernova field, and the pessimists were supernova observers who had a lot of experience in making mistakes.

Either way you approached the problem, whether by building up knowledge of nearby objects or straining to find distant ones, the path was sure to be difficult. Supernovae are rare events, taking place only about once per century in a galaxy. Whether you are looking for distant supernovae to do cosmology or nearby supernovae to learn about supernovae, you have to work hard to find them.

In Boston, at the end of Commonwealth Avenue, the city's grandest boulevard, there is a large red sandstone Viking boat, capped by a noble sculpture of Leif Ericsson, looking westward across the Muddy River toward Fenway Park. The inscription on the front is runic (I guess), but on the back it says, "Leif the Discoverer, son of Erik, who sailed from Iceland and landed on this continent A.D. 1000." There doesn't seem to be any doubt that the Vikings reached North America long before the explorations of Christopher Columbus. But the history of European settlement of North America, written by those who stayed, doesn't have much to do with those Norsemen, who came and left long before the Pilgrims stepped on Plymouth Rock. My third-grade teacher never mentioned Lief Ericsson. The Vikings were too far ahead of their time.

The same thing happened again at the European Southern Observatory, located in the north of Chile, in the mid-1980s. A brave group of optimistic Vikings set out to find type Ia supernovae in clusters of galaxies, with the aim of measuring cosmic deceleration. They were ahead of their time. Although they invented most of the methods later used in high redshift supernova searches, 1980s technology was not quite up to the task of determining cosmic deceleration. Subsequent success builds more on technological change than on brilliant insight. During 1986 and 1987 Leif Hansen, Hans Ulrik Nørgaard-Nielsen, and Henning Jørgensen traveled from Denmark to Chile every month, racking up astonishing amounts of frequent flyer miles. In Chile, they used the Danish 1.5-meter telescope with a 300 × 500-pixel charge-coupled device (CCD) electronic camera to make images of a selected set of galaxy clusters every month.

If a supernova goes off once in a century per galaxy, that's roughly once in 5000 weeks, so if you want to see a nice fresh supernova at its brightest tonight you need to examine several thousand galaxies.[3] The Vikings sought to maximize their chances by looking at galaxy clusters where the number of galaxies in their tiny image of the sky would be well above the average. Also, these were galaxy clusters with known redshifts. If you're looking for supernovae to give you information on cosmology, then you want to search in galaxies at a big enough redshift so that the cosmological effect of deceleration, which would make the supernovae seem a little brighter, would be detectable. By carefully selecting their target clusters, the Danes emphasized galaxies in the optimum redshift range. Also, dense galaxy clusters have lots of elliptical galaxies, the kind with old stars and little dust. Type Ia is the only type of supernova ever found in elliptical galaxies. So the Danes were betting that any supernova they found would be a SN Ia with little dust to dim its light and confuse the analysis.

They came back month after month. A month is the natural rhythm for astronomical observations, since it is the period for the waxing and waning of the moon. Since observing faint supernovae demands dark skies, you generally need to observe within a few days of new moon. I schedule my life around the phases of the moon as diligently as a werewolf because conditions for faint-object observing are best at new moon. You get a chance to do this every 29 days, more or less. By good luck, this is in good accord with the 20 days it takes for a type Ia supernova to rise to maximum light and the two weeks or so it spends within a factor of two of its maximum brightness. Searching more frequently would not help much—you'd see the same object many times. Searching much less frequently than once a month would not be so good either, because then you couldn't tell whether an object you saw for the first time tonight was a fresh one on the way up or a stale one on the way down.

The Danes also pioneered a valuable technique in actually finding the supernovae. Where Charlie Kowal scanned films at Palomar the same way Zwicky had with his gimlet eye 50 years earlier, the Danes used the digital data from their electronic camera to look for supernovae in their computer. They took an image of a galaxy cluster and stored that on disk. While they were exposing on the next cluster, they examined the image of the cluster they had just done, comparing it to a template image of the same cluster from a month ago or a year ago. Instead of comparing the two by eye, they used their computer to subtract the old picture from the new one. By carefully aligning the images, blurring the better one to match the one taken under less good atmospheric conditions, and scaling them so the subtraction makes constant objects disappear, a cluster of hundreds of galaxy images can be simplified to show only the things that have changed from one month to the next. Some of these things are galactic nuclei (with giant black holes!) that vary in brightness, some are asteroids in the solar system, some are cosmic rays at the Earth's surface that make a false signal in the detector, but once in a great while, you find a dot that is present in tonight's image, absent in last month's, and a plausible candidate for a supernova in a distant galaxy.

On 9 August 1988, after many months of searching, the Vikings found what they were looking for: a nice new dot in a galaxy in a cluster at redshift $z = 0.31$. They had arranged with their colleagues

Richard Ellis and Warrick Couch to get a spectrum of their target at the 4-meter Anglo-Australian Telescope and light curves at the 2.5-meter Isaac Newton Telescope in the Canary Islands. They submitted their discovery for announcement in the IAU Circulars, run by Brian Marsden, at the Harvard–Smithsonian Center for Astrophysics. Brian called me. Was this report interesting enough to include?

At first, I was skeptical—I didn't know anything about the Danish search, but the reported supernova was so faint that I didn't think anybody would invest their own telescope time to follow up. The most likely thing was that this was the dim fading tail of an old, dull supernova. But the Danes, quite rightly, insisted to Brian that faint supernovae were precisely their targets and that they had been diligently searching the same fields each month to find them. This supernova was probably faint because it was distant, not because it had faded. Brian had them add more detail to the message, saying clearly that this was the result of a focused search for supernovae in distant clusters of galaxies, so people who read the circular could understand why this faint supernova was worth pursuing.

In the end, the useful observations all came from the arrangements they had made themselves beforehand, including getting a spectrum with the Anglo-Australian Telescope. They wrote up the results for *Nature*, the British science journal. Since most scientists can't penetrate the jargon to read research articles in fields outside their own, *Nature* helps us out by having another scientist write an accompanying "News and Views"—a translation of the most interesting articles into useful and polite literature. Then a biologist can enjoy new developments in astronomy or an astronomer can find out what geologists are doing. *Nature* thought this discovery was worth the ink and asked me to write the "News and Views."

> Although it is appealing to think that supernovae might lead us out of an age of ignorance and belief into an era of measurement and understanding, two observational issues need to be carefully studied before too much faith is placed in this promising approach. First, the homogeneity of the type Ia events is a matter of observation, not of faith, and there are recent examples that show small, but real, differ-

ences among members of this class. . . . . Second, we need to build confidence that the supernovae observed at high redshift are really the same as the supernovae observed nearby.

The skeptical, but not hostile, author continued,

> [A good] approach might be to shore up our knowledge of supernovae from the local neighborhood through intermediate redshift to be certain that any observed effect comes from space curvature and not from a changing population of supernovae.[4]

The Danes were too far ahead of their time. Their telescope was small, so it took an hour to take a single image. Their detector was small, so the area of the sky they could search was tiny. Working slowly on tiny fields meant that even with the best weather and best technique, the rate of discoveries was very low. After finding just one SN Ia (and another that was probably a SN II) in two years, they decided to give up. Ironically, they found another good candidate on the very last night of operation, but decided not to report it, since there was no chance to follow up to get the light curve. Like the cod-dryers of Vinland, though they *were* the pioneers, this group of Norsemen was too early to be part of later developments.

The real weakness of the Danish search was that the rate of discoveries was so low that you could not plan the follow-up observations with any certainty. Telescopes at big observatories are often scheduled six months in advance. The observers on a given night are not likely to think your work is more important than the project they have waited six months and traveled 8000 miles to do on their three nights. On the other hand, if you want to schedule the follow-up of supernovae, you need to convince the Time Allocation Committee that, despite their infrequent eruption in a single galaxy, your search is so powerful you are sure to have some supernovae to follow.

This problem was solved by the Calán/Tololo supernova search carried out at Cerro Tololo between June 1990 and November 1993. At the Supernova Workshop that Stan Woosley organized in Santa Cruz in July 1989, Mark Phillips, then a Tololo staff member, asked if I thought it would be worth doing a supernova search. Mark, a tall, gangly Californian, had led the Cerro Tololo basketball squad to the La Serena, Chile, city championship. He had worked on the emission from gas swirling down into black holes in the centers of galaxies. He was ready for something new. Golf and supernovae.

"Only if you can follow them up," I replied, recalling my little orphans SN 1971M and N. Mark thought about it, and got together with José Maza, Mario Hamuy, and others from the University of Chile's Cerro Calán Observatory to find supernovae the old-fashioned way by searching photographic plates taken at the wide-field Schmidt telescope on Cerro Tololo. This was the University of Michigan's Heber D. Curtis Telescope, named after Harlow Shapley's worthy opponent at the 1920 debate on the nature of the spiral nebulae. It was Curtis who imagined that "a division [of novae] into two classes is not impossible," with an extra-bright kind of nova required in the spiral nebulae if they were really at large distances. It seemed fitting for this telescope to be the vehicle for studying supernovae.

The Calán team would "blink" the plates, using an optical contraption that presents first one and then the other image to a trained eye. Something new—a candidate supernova—would blink on and off. Nick Suntzeff, one of the world's experts on doing brightness measurements right, was part of the team. Mark and Nick worked out an arrangement so that visitors to Cerro Tololo knew in advance they might be asked to give up an hour to help with supernova observations. In addition, regular blocks of time were allocated to the supernova team because they were certain to have some fresh supernovae every month. In exchange, Mark generously offered to make every contributor an author of the subsequent papers.

This plan worked well: every month, the Curtis Schmidt telescope surveyed a chunk of the sky. Although the detectors were photographic plates, which are much less efficient than electronic detectors, the area on the sky was large. The plates were developed at Tololo, then shipped down the Pan-American Highway to Santiago on the bus. Although blinking plates was tedious, the Calán team was experienced and willing. And though the supernovae discovered in the Calán/Tololo search were not distant enough to measure cosmic deceleration, they were far enough for redshifts to give good distances and they were well observed in a consistent way by an expert team. This program built up the data that made the next step forward possible. In three years, they found 49 supernovae, and followed up 31, eventually creating a breakthrough that made the cosmological use of supernovae a reality. But it takes time to get a big survey finished.

Meanwhile, the evidence for inhomogeneity in SN Ia was growing. In 1991 two unusual supernovae, SN 1991T and SN 1991bg, were discovered, which strengthened the case for real differences among SN Ia. SN 1991T appeared to be the brightest SN Ia known and its spectrum was subtly different from those of normal SN Ia. At the other extreme, we observed SN 1991bg, which was apparently one of the lowest luminosity SN Ia, with some clear differences from the standard SN Ia spectrum in the first week after maximum brightness. These were very well observed objects, intensively studied by Alex Filippenko's team at the Lick Observatory near San José, by our Center for Astrophysics group at the Whipple Observatory near Tucson, and by Cerro Tololo. There couldn't be any doubt that SN Ia were not all the same. The differences in the spectra were subtle, but the differences in the light output were not so small—SN 1991bg appeared 10 times dimmer than an earlier SN Ia in the same galaxy. There would be no easy way to measure the 25 percent effects due to cosmic deceleration if the supernovae themselves were introducing 1000 percent effects!

People who were depending on SN Ia to be standard candles could not ignore this evidence. As this lesson seeped into the consciousness of astronomers, some lost their faith. Sidney van den Bergh, long an expert on supernovae and an independent voice on the Hubble constant, abandoned hope, saying, "supernovae of type Ia have a large luminosity dispersion at maximum light, and may therefore not be good standard candles." Some called it heresy. Tammann, for example, emphasized how good the uniformity was, if you could filter out the brightest and the dimmest by their colors and spectra. Others venturing into the field didn't worry too much about these astronomical details. The fledgling Supernova Cosmology Project at Lawrence Berkeley Lab, led by Carl Pennypacker and

Saul Perlmutter, concentrated on developing methods for finding distant supernovae. But by 1992, it was clear to those of us working on these objects that the problem was serious and real. If somebody didn't find a way to deal with the fact that some SN Ia were 10 times brighter than others, there was not going to be much cosmology done with exploding white dwarfs.

The first step toward a solution was not far off. Mark Phillips and the CTIO group had observed a strange supernova in the galaxy Centaurus A (NGC 5128) back in 1986. SN 1986G was weird. The spectrum looked like a SN Ia, but the light curve declined much more rapidly than SN 1972E or the other well-observed objects that made up Bruno Leibundgut's template. SN 1986G was fainter than other supernovae would be at the distance of NGC 5128. Something was either wrong or new. The things that could be wrong were the distance to NGC 5128, which is too nearby for the redshift to be a reliable guide, or the amount of dust absorption. The thing that could be new was that type Ia supernovae were not as uniform as claimed. Mark looked into this. In 1992, he plotted the luminosity of several supernovae, including SN 1986G, based on his best estimate of the supernova distance, against the amount by which it declined in brightness during the weeks after maximum light. There were real differences among the SN Ia. The intrinsically brightest supernovae declined the slowest. The distance estimates in Mark's paper were a patchwork and the number of extra bright and extra dim supernovae was pathetically small. This result could have been wrong. There was a long tradition of getting things wrong about supernovae and their decline rates.

There had been previous claims of inhomogeneity among the type Ia supernovae. In 1973, Roberto Barbon and the Italian supernova group in Padova had compiled all the old photographic data and pointed out that the light curves were not identical. They suggested that there were "fast" SN Ia, which declined rapidly, and "slow" SN Ia, which declined slowly after maximum light. At that time, Gustav Tammann stoutly defended the homogeneity of SN Ia against the "fast" and "slow" heresy. He pointed out that in any sample random Gaussian errors will create a fastest and a slowest light curve, but that doesn't necessarily mean the supernovae fall into two classes. It was possible, Tammann claimed, that the supernovae were all identical, but that measuring errors had produced a distribution of decline rates, and that was the origin of what Barbon was seeing. Barbon could be credited with discovering the importance of supernova decline rates, but that wouldn't quite be right. In their data, Barbon and his colleagues found no clear connection between the intrinsic brightness of a supernova and its decline rate—but the data hinted that the fast supernovae were the brightest, while the slow supernovae were the dim ones, exactly opposite to what Phillips concluded twenty years later. When Yuri Pskovskii of Moscow examined this question in 1977, he found a relation like the one Mark Phillips eventually found, but it was not widely applied.

The problem with Barbon's work was not the analysis, but the data. They were using inhomogeneous sets of data patched together from many workers dating back to Baade. And the light curves were almost all extracted from photographs. Since the light from a supernova is usually on top of the light from the galaxy that hosts it, it is a delicate task to subtract the galaxy light to get the brightness of the supernova alone. Photographic plates are notoriously tricky in this regard: if you add together the light from a galaxy and a supernova, the resulting blackening on the photographic emulsion is not the same as if you added up the effect of the galaxy alone and the effect of the star alone.

By the 1980s silicon diode arrays, CCDs, were standard equipment at Cerro Tololo, at our Whipple Observatory in Arizona, and at many other places around the world. As anybody who has used a camcorder knows, CCDs are much more efficient and work without a flash in dim light. More importantly for astronomy, where the aim is to extract a measurement of a supernova on top of bright sky light plus galaxy light, the detectors are linear. That means the sum of the signals from a star alone and the galaxy background by itself was equal to the signal from the sum of the galaxy light and the supernova light. This linear relation between the light coming in and the signal coming out makes the process of extracting supernova light curves from the data much simpler for CCDs. Simple pictures are best, and much more likely to give reliable answers.

The Calán/Tololo search was carried out on photographic plates, but all the measurements of supernova brightness were done with CCDs. Slowly, carefully, the data were reduced by Mario Hamuy, Nick Suntzeff, and the team in Chile. Photometry, the science of measuring the brightness of things, is deceptively difficult. Even without making any overt errors, small things can add up to make your data worthless. Photometrists know this and expert photometrists know this best. It makes them a little dour.

Nick Suntzeff is the photometrists' photometrist. And Nick is bit of a pessimist, like Eeyore in *Winnie the Pooh*. He worries that the filters in the camera might not be right, that the standard stars we use to calibrate the supernovae might not be as well known as everyone assumed, that the weather wasn't as good as people thought, and therefore the results might not be as good as claimed. You want Nick on your team, because Nick keeps you from making careless errors, because Nick doesn't assume you've done things right, and especially because when Nick finally says the data are OK, the data really are OK. Master craftsmen work slowly, and the results from the Calán/Tololo survey took time to reach perfection.

Meanwhile, in 1992, Mark Phillips was out on a limb, using an inhomogeneous data set and a variety of ways to estimate supernova distances. He could have been wrong about the luminosity–decline rate relation. But as data from the Calán/Tololo Survey and other sources trickled in, the luminosity–decline relation looked better and better. The Calán/Tololo Survey found supernovae in galaxies at moderate distances—typically 600 million light-years. That is close enough for the follow-up measurements to be feasible with medium-sized telescopes, but far enough for the redshift to stand in well for the distance.

At the same time, we were building up our own efforts at the Center for Astrophysics (CFA). Bruno Leibundgut came from Basel as a postdoc to help clean up my backlog of supernova data. Piles of magnetic tapes were growing like stalagmites, clogging my office. Bruno carted them away and started analyzing the data. Ron Eastman, who had come with me from Michigan to Harvard, was finishing his theoretical thesis on the atmospheres of type II supernovae. When Bruno went off to Berkeley to work with Alex Filippenko,

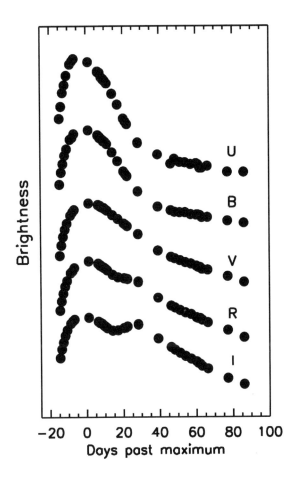

Figure 9.1. **The lightcurve of a type Ia supernova.** Measurements made at the Smithsonian's Whipple Observatory in Arizona of SN 2001V, discovered by Perry Berlind of the observatory staff. Measurements were made in five different colors, displaced here for clarity. The slope of the declining light curve in the "B" (for blue) filter contains powerful information about the supernova's true brightness. Observations in other filters contribute to the precision of the luminosity determination and also tell the amount of reddening by dust. Courtesy of Saurabh Jha, Kaisey Mandel, Tom Matheson; Harvard-Smithsonian Center for Astrophysics.

Pilar Ruiz-Lapuente came from Barcelona as a postdoc. David Jeffery came to work on the theory of supernova spectra, and he was puzzling out what made SN 1991T and SN 1991bg different. With support from NASA and the National Science Foundation, I was able to hire Pete Challis to work on Space Telescope data and to help with the supernova observations we were doing in Arizona. Brian Schmidt was working on his thesis on type II supernovae, Adam Riess signed up to work on his Ph.D. with me, and energetic postdocs Peter Höflich and Phil Pinto were working independently on the theory of supernova spectra. In a small field, this was a big team.

On Fridays, everybody on the CfA team who wasn't off observing would go to lunch. This was my way of keeping up with what everybody was doing. As you get to be the Professor, sometimes your personal "progress" for the week is finishing a grant proposal, refereeing a paper, and teaching classes. Those are the weeks when your most effective contribution to science may be to pick up the check at lunch. Lunch was a good place for the beginning students to hear what others were doing and to learn to describe, briefly, their own progress.

My rule was, you had to say what you had done in the past week, but you were only allowed one sheet of paper, perhaps a graph or a picture, to supplement your description. This kept the conversation going and prevented fast-flipping viewgraphs from substituting for scientific communication. It was also a lesson in cultural styles. As Americans, we were ready to eat, listen between bites, and go. But Pilar showed us another way. Just as I was on the brink of getting the check, she would ask for "a coffee," and we would then learn something of the Barcelona way of life, letting the lunchtime stretch just a little into the afternoon, letting the conversation drift just a little beyond interstellar violence and cosmic expansion. But not much. We usually stuck to the subject, though one time Peter Höflich staggered us all by flipping out baby pictures of his day-old daughter. We didn't even know he had a girlfriend.

We were in close touch with the CTIO group, collaborating with Nick and Mark on studies of SN 1987A and other supernovae with the Hubble Space Telescope, sharing and comparing data. While

the principal target of the Calán/Tololo search was SN Ia for the
Hubble constant and cosmology, you don't know what type of supernova you have discovered until you take its spectrum. Along
with the type I's they were looking for, like fishermen with a broad
net, the Calán/Tololo patrol was also trawling up some SN II. My
student, Brian Schmidt, visited CTIO in 1991 and instigated an active collaboration that helped him complete his thesis, using some
of the type II supernovae from the Calán/Tololo search. Brian's plan
was to gather up the SN II observations, which were piling up on
the floor in La Serena as a by-product of the Calán/Tololo search.
If Brian could measure the supernova colors and the speed of the
atmospheric expansion from the light curves and spectra, he could
use them to get a better value of the Hubble constant from the expanding photosphere method using some of the theory that Ron
Eastman was developing.

In 1993 Brian Schmidt finished his Ph.D. at Harvard. Although
we usually liked to push the fledglings out of the nest, Brian was
so extraordinary he won one of the competitive postdoc jobs at the
Center for Astrophysics. This gave him the chance to step out as an
independent worker and to double his salary without moving. Brian
decided to visit CTIO, and talked with Nick Suntzeff and with Mark
Phillips about what should come next in studying supernovae.

Meanwhile, having begun in 1986, a serious effort to find and
study supernovae was developing at Berkeley. The combination
of the Berkeley Astronomy Department with Alex Filippenko, the
Berkeley Physics Department with Rich Muller, and the Lawrence
Berkeley Laboratory (LBL), including Carl Pennypacker and later
Saul Perlmutter, had been working on various aspects of supernova
science. Muller, a brilliantly inventive physicist, decided to turn the
process of finding supernovae from a craft into an industry.

Years earlier, in New Mexico, Stirling Colgate had cobbled together a supernova search telescope from a surplus Nike missile
turret and the primitive computers of the 1970s. He built an automated system that could point at galaxies one after another without
human intervention, taking an image in a few seconds. Computer
software would then examine the image, and sound an alarm when
a new object was detected. But Stirling Colgate was not quite the

Leif Ericsson of supernova searching. Stirling was so far ahead of his time in so many technical areas that he never got all the pieces working together long enough to find even one supernova. He never got to Vinland.

Rich Muller knew that technology had evolved, and, after being rebuffed by the Air Force in a proposal to use their tracking telescopes at Kwajalein atoll to look for supernovae in their classified data stream, he inspired the effort to get Berkeley's 30-inch telescope east of the Berkeley Hills operating in the way that Colgate had envisioned.[5]

After some agony, it began to work, and the Berkeley Automatic Supernova Search Team began to find supernovae in 1986. What was especially good about this approach was that you could keep careful records of the galaxies searched and use that information to figure out the rates of supernovae in various galaxy types. Best of all, if you adjusted the observing cadence of the search, you could maximize your chances of finding supernovae on the way up, before they reached their maximum brightness. Getting the supernova search telescope working took technical innovation, but building up results on rates took patience and dedication to the subject.

Rich Muller's brain was too effervescent to plod. Also at Berkeley, physicist Luis Alvarez and his geologist son Walter were beginning to piece together evidence that the Earth had been bombarded by asteroids. About 65 million years ago, one of these killer rocks had whacked into the Yucatan, shrouding the Earth in dust, making life stressful, perhaps to the point of extinction, for the dinosaurs.[6] Further investigation of the cratering history of the Earth suggested that episodes of bombardment were periodic, recurring roughly every 26 million years. One hypothesis was that the sun had a distant companion—a dim star 160 times farther from the sun than Pluto, which slowly made its way around an elliptical orbit. Every 26 million years, according to this idea, there would be a rain of doom as the Nemesis star nudged rocks in the outer solar system into orbits that would bombard the Earth. This idea was so interesting to Muller that the automated telescope was partly diverted from supernovae and the fate of the universe into searching for Nemesis and the fate of life on Earth. This isn't the choice I would have made, but you can see that an Earthling might be interested. Muller didn't find Nemesis, though it may yet be lurking out there. Or perhaps there is some other cause for the periodic bombardment. Or perhaps the geological evidence for periodicity is not as strong as it seemed at first.

In any case, the idea that supernovae were interesting and possibly a route to learning something about the fate of the universe remained alive. LBL had working software that could find a new supernova in an image of a galaxy and had shown that this system could work on individual galaxy images with a small telescope. It wasn't that great a leap to think that similar software could work on an image that contained many galaxies from a large telescope as the Danes were doing at the European Southern Observatory. LBL worked out a deal with the Anglo-Australian Telescope to build a big, very fast CCD camera that could get the data for this program by installing it on that 4-meter telescope in exchange for time to use it to hunt for supernovae. The optical design was very daring. But the instrument never worked satisfactorily and that LBL effort never reported a supernova.

In 1989, UC Berkeley won a national competition sponsored by the National Science Foundation to fund a new science center to address the question of dark matter in the universe. The Center for Particle Astrophysics was ably led by Bernard Sadoulet, formerly a lieutenant of Carlo Rubbia at CERN, the European accelerator near Geneva. The idea of the center was to learn about dark matter in a large number of ways. Their artfully designed T-shirt said, "If it isn't dark, it doesn't matter." Sadoulet himself would take the direct approach, building laboratory detectors to see if dark matter particles were drifting through the room. Another group would look for the signature of dark matter in the microwave background. Theorists would weave all of this together into a coherent story for the evolution of a dark matter universe. And supernovae would be used to measure the amount of dark matter by detecting cosmic deceleration. If $\Omega = 1$, then at redshift 0.5, the supernovae should appear 25% brighter than otherwise. The Supernova Cosmology Project (SCP) was going to make that measurement. LBL was experienced in supernova detection software, had capabilities in advanced instrumentation, and, as experimental physicists, understood the analysis of subtle data sets. They would lead the way, with help from Alex Filippenko in the Berkeley astronomy department, who joined the project in 1993.

To stimulate this enterprise, in 1989 they organized a symposium in Berkeley to bring together all of these strands in modern astrophysics. I gave a talk on "Attacking H_0 and Ω with Supernovae." Despite this bellicose title, my conclusion, based on the Danish work, was timid: "these pioneering observations point out the possibility of making progress on the cosmological problem from diligent observation." In fact, I thought that the scatter among SN Ia was so large, that since the number of supernovae you need increases as the square of the scatter, you would need so much diligent observation that we should build a special 4-meter telescope with supernovae and Ω in mind at a cost of $10 million. The cheaper path would be to understand supernovae better.

When the National Science Foundation established the Center for Particle Astrophysics, Bernard Sadoulet asked me to serve on their External Advisory Board, which was supposed to help evaluate the many activities of the Center and advise him on choices he had to make. The supernova team was having trouble. After the dead end at the Anglo-Australian Telescope, they did not have a working camera on a telescope where they had guaranteed access. They were going to have to compete with the rest of the astronomical community for time at Kitt Peak or Cerro Tololo. But their credibility in that community was not the best after the 1987A pulsar report, the diversion of the supernova search to Nemesis, and the camera's failure in Australia. Although they had invested serious effort in the supernova-finding software, they hadn't yet found any distant supernovae, so time-allocation committees were reluctant to give them scarce telescope time to carry out a search. If they didn't search, they weren't going to find any supernovae. To get out of this catch-22, Bernard convened an outside committee.

That group proposed putting Perlmutter in charge. Although he was quite junior, Saul was very determined, had good judgment about what was most important, and made a forceful spokesman for the project. Maybe he could convince people to give them the telescope time they needed. They also proposed more money and a program to acquire large CCD detectors to put in a camera on a British telescope in the Canary Islands, in exchange for guaranteed time. Though the outcome was good for them, the SCP didn't like undergoing all these reviews.

While the LBL crew was struggling with all this, I would breeze in periodically for the External Advisory Board meeting. As I recall, I emphasized three things. One was that photometry was hard, and they should not underestimate the difficulty of making accurate measurements of faint objects. Another was that there was growing evidence that SN Ia were not all alike, and they should pay close attention to this work. And finally, there was a history to this subject, and the lesson of history was to watch out for dust. If they didn't make measurements to determine reddening, there would be problems later with the interpretation. Alex Filippenko, from the Berkeley astronomy department, gave them similar warnings. Nobody at LBL really wanted to hear all these cautions—they had their hands full figuring out how to find distant supernovae. I realized just how unwelcome these suggestions were when SCP member Gerson Goldhaber later described this period by saying, "Bob Kirshner was pooh-poohing our research every step of the way; he said the approach would never work."[7]

Finally, the Berkeley team got a break. In 1992, they found SN 1992bi using the 2.5-meter Isaac Newton Telescope in the Canary Islands. Because of this discovery, they were successful with the Kitt Peak time allocation system and won time to search with the observatory's standard CCD camera at the 4-meter. By 1994, they had six objects. Saul proved to be a master at getting other people to observe their discoveries. He would track you down in the control room of a telescope anywhere in the world, impress on you how important his work was, and try to convince you that it was more important to observe his supernova tonight than your own program. It was a tough sell, but Saul was relentless. People might roll their eyes, but they would take the data he wanted.

Of course, I had been on the other end of these exchanges myself many times, diplomatically hoping to get an observer to take some unscheduled data of a particularly interesting object. Supernovae are different from most astronomical objects. Most objects will be there next year, so if you don't do them tonight, you can do them next year. But with supernovae, if you don't act now, the chance will pass, and you will lose them forever. It adds drama to observing. The quid pro quo, most effectively instituted at CTIO, but which I had also implemented at the CfA, was first to get the authority to butt in from the Time Allocation Committee and then to give credit to everybody who contributed data—including them as authors of the resulting scientific publication.

In response to his phone calls, I observed Saul's objects myself, getting a spectrum of SN 1994G at the MMT in Arizona. At that point this was the best spectrum ever taken of a high-redshift supernova. I shared my data with SCP. I was surprised when they presented it at the next advisory board meeting I attended as "a spectrum we have obtained."

In August 1993 the LBL team submitted their first scientific result for publication in *The Astrophysical Journal Letters*, a description of their work on SN 1992bi, a supernova in a galaxy at redshift 0.46. Publication in a reputable journal is the moment when a scientific team gets credit for their work. It is important, even in a world where electronic preprints and meeting abstracts are lesser forms of telling the world what you've done. In astronomy, as in most academic fields, the editors of a journal send a paper to a "competent referee" who is supposed to read it carefully, offer comments or suggestions for improvement, and advise the editor whether the paper is suitable for publication in that journal. Referee's reports in astronomy are usually anonymous to avoid retribution for frankness. A typical referee's report might point out omissions ("this paper contains too few references to the work of the anonymous referee"), errors ("the statement at the end of the paragraph is wrong—a standard candle appears dimmer in an empty universe"), as well as offering a judgment ("this paper is both novel and correct. Unfortunately, the parts that are correct are not novel and those that are novel are not correct").

The Astrophysical Journal Letters is a U.S. journal with high standards—to get in, an article needs to be short (four pages, maximum) and very interesting. Since I wasn't an author and knew something about supernovae, the editor sent this paper to me. At first, I was delighted. After all, this is a paper I would read carefully in any case. Then I read it and was not so delighted. It was short and interesting, but a reader couldn't tell if it was right. It seemed to minimize three things. That photometry is hard. That SN Ia are not all alike. And what about dust? Because of the way they observed this object, the SCP did not have any information on the color of this supernova, so they had no way to say anything about the effects of dust, which could easily be as big as the effects of cosmology. Maybe the supernova was much brighter due to a decelerating universe, but this was balanced out by dust absorption. There was no way to tell. Since the true brightness of the supernova was the central point of the paper, I thought they had a real problem.

What to do? On one hand, you owe the journal a frank appraisal (especially if the journal editor has his office four hallways away!); on the other, you hate to make life harder for people who are busting a gut to do something important. I sent a very detailed report, recommending a lot of changes before publication. The authors revised the text, but I still wasn't convinced they had dealt with the central issues. Maybe it wasn't possible in the four-page format of *The Astrophysical Journal Letters,* and they should consider writing the *War and Peace*–length version for another journal. Authors don't have to accept the verdict of a single referee, who might be pig-headed. They can ask for another. Which in this case, they did. Journal editors figure the chance of getting two village idiots in a row is small. The second referee wrote a long report, generally concurring with mine and suggesting a major change in the paper's emphasis. Then the first referee gets to see what the second referee said. Neither of us recommended publishing the paper in its present form.

The editors, sensibly, err on the side of caution, not willing to publish something until people more-or-less agree and somebody says, "this should be published." The authors can revise their paper to take the referees' comments into account. All of this back-and-forth takes time. Their paper, with more modest claims about cosmology, appeared in the 20 February 1995 issue of *The Astrophysical Journal Letters*.

While the Supernova Cosmology Project was getting underway at LBL, the Calán/Tololo Team had begun to crack open the problem of what to do about the big difference in the brightnesses of SN Ia like SN 1991T and SN 1991bg. Supernovae are not all alike, but there is a way to deal with it. Mario Hamuy was the lead author of a paper from the Calán/Tololo team that took the nugget of an idea from Mark Phillips and turned it into a real solution to this puzzle. Mario's paper showed that Mark was right: the slow declining supernovae are the bright ones and the fast decliners are the dim ones. If you measure how fast a type Ia supernova fades after it reaches maximum light, you learn whether it is on high beams or low. If you know that, you won't foolishly assign it the wrong distance.

This result was very important for the program of using supernovae to measure cosmic deceleration. Instead of a big range of brightness that causes big distance errors, the use of the light curve shape for SN Ia decreased the error in distance for a single measurement to about 7 percent. This moved the problem of measuring Ω from a major undertaking requiring its own 4-meter telescope to a reasonable observing program that could probably be done by a determined group with existing facilities in the span of a single graduate student's thesis.

At the same time, the Cerro Tololo team let us at the CfA see some of their light curves. Combined with our own data from Arizona, we then had a good set of light curves to look at the connection between the decline rate and the true brightness. Graduate student Adam Riess, with mathematical inspiration from Bill Press, another professor at Harvard, and astronomical advice from me, developed an alternative way to use the light curves to determine the intrinsic brightness of supernovae. Adam started from Bruno Leibundgut's template light curve, then examined how the light curves of brighter or dimmer supernovae differed from the template. It was a neat piece of work that gave results as good as the CTIO group's method. These methods also gave quantitative estimates of just how good the distance measurement was for each supernova. Knowing your errors is very helpful in knowing how far to trust the conclusions. Some objects are observed many times and the light curve is great; others have spotty data due to weather, failure to twist the observer's arm, or other causes. It matters whether you know the distance to a supernova well or poorly when you are trying to measure cosmic deceleration. The "light-curve shape" method (LCS—which we thought was droll in the year of the baseball strike when there was no League Championship Series) told us the sigma: how trustworthy each distance measurement was and how much it could add to a measurement of the cosmic deceleration.

Adam didn't stop there. I was worried about dust. If supernovae nearby were found more easily in dusty regions than the supernovae in distant galaxies, then you might end up with local supernovae being dimmed, distant supernovae appearing brighter, and a false signal for cosmic deceleration that was due only to failure to account for absorption by dust. How embarrassing would that be?

Adam and I noticed a very troubling property of the data for local supernovae. If you took the data for most of the supernovae from the time of Baade and Zwicky, and assumed they had the same intrinsic color at maximum light, then you could use the observed color to estimate how much they had been affected by dust. For example, if you knew the real color of a supernova was blue, but you measured yellow or red, you would know that wicked dust was between you and the supernova, dimming it and making it appear redder. The trouble with this simple picture was that when you took a sample of data and corrected in this way, instead of decreasing the scatter, the scatter of the points got bigger. This is Nature's way of telling you that you've done something stupid and that instead of correcting for reddening, you are somehow making things worse.

The solution was not so complex. What if, instead of assuming that all supernovae had the same color, you assumed that the color might depend on the brightness? After all, the light curves of the bright ones declined more slowly and the spectrum was a little different, so why couldn't the colors be different, too? The bright ones might be blue and the dim ones red. In fact, if the spectrum differences were caused by temperature differences (which is usually the case), then you'd expect blue color, high temperature, and an extra-bright supernova like SN 1991T to show a spectrum that was a little different from a red, cool, dim one like SN 1991bg. Adam was able to solve separately for the light-curve *shape* as observed in one filter (which tells you the true brightness and the intrinsic color) and the measured *color* using another filter, which tells you how much the supernova light has been reddened by dust. And if you made measurements through more filters, each sampling the light in a different color, you learned still more about the true distance of the supernova.

With this in mind, all the new data we had been taking in the CfA sample and all the new data from Cerro Tololo were observations taken in several colors, ranging from the ultraviolet to blue to green to red out into the infrared where the CCD detectors work, but your eyes don't. We called the new and improved method MLCS (for "multicolor light-curve shape"). The Calán/Tololo crew developed an independent method that also allowed a measurement of both the distance and the dust absorption. Both groups had figured out how to use the light curve information to make type Ia supernovae into the best distance-measuring tools for cosmology.

The results were excellent. When we used MLCS on the data for nearby SN Ia, we could reduce the scatter from about 40% (if you assumed they were identical standard candles) down to less than 15% by using information about the light-curve shape and the color to see which were bright and which were dim. Using the methods developed by Mark Phillips and his collaborators worked equally well. Since random Gaussian errors get driven down by the square root of the number of measurements, the number of supernovae you need to see the difference between an $\Omega = 1$ universe and an $\Omega = 0$ universe depends on the *square* of the error associated with each data point. Reducing the scatter in brightness for a sample of SN Ia from 40% to under 15%, about a factor of 3, meant you could make the cosmological measurement nine times faster!

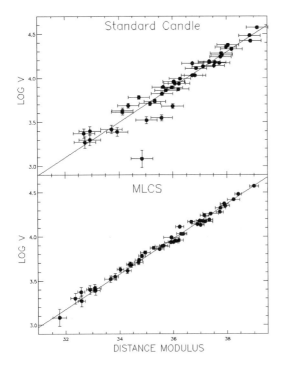

Figure 9.2. **The good effect of using the light curve shape.** The top panel shows a Hubble diagram of the redshift versus the distance (in astronomers' units). If you assumed all type I supernovae were identical and you judged the distance from the apparent brightness, you would get the Hubble diagram in the upper panel. When a supernova is intrinsically dim, this approach mistakenly assigns it an extra-large distance. The lower panel shows the Hubble diagram after correcting for light curve shape and reddening using the MLCS. The improvement is dramatic. The 1σ error drops from about 15 percent to 7 percent error in the distance. This means each measurement becomes four times as useful. Courtesy of Adam Riess, Harvard-Smithsonian Center for Astrophysics.

To me, this vindicated our long-term strategy of concentrating on learning the properties of nearby supernovae before attacking the cosmological problem. Even though we'd been paddling around in the shallow end of the pool, while the SCP had been diving in the deep end, learning how to find distant supernovae, now we were ready to swim the English Channel. In fact, if you believed the errors from the MLCS, you would get a strong hint about Ω from just one supernova. The difference in apparent brightness at $z = 0.5$ between an $\Omega = 1$ world and an $\Omega = 0$ world was 25 percent. Our uncertainty, if we had light-curve measurements as good as the ones from the Calán/Tololo sample or the CfA data, was just 15 percent. Of course, we couldn't expect the data on faint and distant objects to be quite as good, and you wouldn't dare to draw a definitive cosmological conclusion from only one object, no matter what your sigma said, but it meant that reasonable data on a handful of objects at redshift of 0.5 could show the fate of the universe. That seemed worth doing.

The LBL team had a handful of supernovae by 1994, but I was convinced they hadn't yet learned anything about cosmology. The SCP data of SN1992bi were, unfortunately, taken in only one filter. With data in one filter there was no way to tell whether dust had dimmed the supernova. So there was no way to separate the effect of dust from the effect of cosmology. You simply could not say how much the acceleration or deceleration of the universe had affected that supernova's brightness because they hadn't gathered the essential data. We had developed the tools to make that measurement right, and we knew exactly what needed to be done.

At this point in 1994, Brian Schmidt returned from his trip to Chile. He and Nick Suntzeff had been talking. There was enough progress on the Calán/Tololo project to see that they were also going to solve the linked problems of supernova brightness and absorption by dust using measurements in several filters. And Nick, being Nick, wasn't convinced the photometry of the Supernova Cosmology Project was up to his own standards of excellence. If the CfA and Tololo worked together with our friends in Europe as the "high-z supernova search team" we surely could do this cosmological problem correctly. "z" is the astronomer's shorthand symbol for "redshift," so "high-z" meant we were searching for very distant supernovae that could tell us something about cosmic deceleration.

Except for one thing. We hadn't found any supernovae at the distances where deceleration would show up. The Calán/Tololo search was photographic—that was yesterday's technology and couldn't be extended to higher redshift. LBL had been working on the problem of automated detection of supernovae since 1986. They had invested many years in building the software for their present system. Their team included highly experienced experimental physicists who were wizards at separating signal from noise in vast data sets. Was it realistic to think we could catch up?

"I think it will take a month," Brian said.

He quickly sketched how we could combine some software packages astronomers used every day to align the new data with the old and subtract it to show only the objects that had changed. We didn't have to reinvent the wheel as the LBL team had done: we could pick up some old wheels at the swap meet.

We applied for time in the first half of 1995 to search for supernovae at Cerro Tololo, using the 4-meter Blanco telescope. We convinced the Time Allocation Committee that our methods were good enough to find distant supernovae, and the enterprise was worthwhile. We were assigned three observing runs with two nights each over three dark runs in February and March, when the moon was down and faint galaxies can be seen. Bruno Leibundgut and Jason Spyromillio applied for time at the European Southern Observatory (ESO) to follow up the flood of distant supernovae we were anticipating. Except ESO has a different calendar for assigning time and we missed the deadline to apply for the first quarter. We were disappointed to be assigned time starting in the second quarter of 1995 to get spectra and light curves at ESO.

Although we had convinced the Time Allocation Committee, Mother Nature had not read our persuasive prose. The first two runs at Tololo were not good and we did not find a single supernova. On the last night of the last run, 30 March 1995, Mark Phillips finally struck gold—a good candidate for a type Ia supernova. While Brian's software had some bugs ("please call them features"), it worked. Bruno Leibundgut carried the finding charts up the Pan-

American highway to ESO's La Silla observatory where he and Jason used the New Technology Telescope to get images and spectra on 2 April. Because they had missed the filing deadline, their observing time had been pushed back just enough to save the whole enterprise from a terrific flop. The spectrum from ESO showed this was a genuine type Ia at redshift z = 0.479, the highest yet observed for a supernova. We announced SN 1995K in IAU Circular 6160. By the skin of our teeth, we were in the game.

But our high-z supernova search was far behind the LBL team. Taking into account the observing runs we had scheduled and the time it takes to completely process and calibrate the data, I figured it would be the middle of 1997 before we would have any results worth talking about. In June 1996, Princeton University celebrated its 250th birthday. Part of the self-congratulatory fun was a meeting called "Critical Dialogs in Cosmology." Princeton University has played a central role in the development of astronomy and of physics and of the combination of the two into modern astrophysics. And, for those without very sharp knowledge of institutional geography, the formidable presence of the Institute for Advanced Study, where Einstein worked, and where John Bahcall has built a temple of excellence for postdoctoral scholars in astrophysics, blurs into the luminous aura of the university. Plus they have a mutant race of black squirrels on campus, so it's always worth taking Exit 9 at East Brunswick on the New Jersey Turnpike.[8]

One of the arenas for "critical dialog" was the status of cosmological dark matter. Was Ω = 1? The meeting organizers opted for the dialectic—they decided to have a debate. As always in science, debates, polls, and opinion are less important than data. One good measurement is worth a thousand metaphors: as a nail is worth a thousand paperclips. But when a subject is murky, with conflicting claims that can't all be true, a debate can at least illuminate our ignorance.

The preponderance of evidence, based on the motions of galaxies in clusters and similar measurements of the mass of clusters, favored a low value of Ω. A good bet was that Ω is 0.3 ± 0.1. That's 7 sigma from Ω = 1. On the other hand, theoretical elegance favors Ω = 1, and when the data are not conclusive, esthetics have some weight. To have a debate, somebody has to take each side. Having no data, I got to moderate. We heard the conventional view that $\Omega = 0.3$, based on the usual evidence from mass associated with galaxies. On the other hand, in the summer of 1996, there was some observational evidence presented for the view that data, not just theory, favored $\Omega = 1$. Avishai Dekel, a former Israeli tank commander, argued forcefully, as tank commanders will, that his method of measuring galaxy motions and inferring the mass that caused those motions pointed in the direction of $\Omega = 1$. I then turned to Saul Perlmutter, who presented the preliminary results of the Supernova Cosmology Project. Saul showed a Hubble diagram with seven supernovae at redshifts where cosmological deceleration would be important. If $\Omega = 1$ then the distance to those redshifts would be a little smaller than otherwise and the supernovae would appear a little brighter. And, according to Saul, that's what the first bit of supernova data indicated—a decelerating universe with Ω not yet well determined, but in best agreement with $\Omega = 1$.

At the coffee break, people asked me what I thought. Since I had nothing useful to say, I was polite. I said these were tough measurements: photometry is deceptively difficult, type Ia supernovae are not all alike, and you need a way to deal with dust. Maybe this wasn't the last word. Our high-z team was also working hard and would make an independent measurement. That's what I said. What I thought was, "Maybe we're too late."

10

getting it right

At the Princeton meeting in the summer of 1996, Saul Perlmutter dropped a bombshell right at the epicenter of cosmology. The supernova evidence accumulated by his group at Lawrence Berkeley Lab favored a universe that was decelerating due to dark matter, with Ω near one. Our high-z team didn't have much to say because we didn't have any results. We had methods that we thought were pretty good, we had found some supernovae and we had some data in the pipeline, but we didn't have our own Hubble diagram to compare with Saul's.

At that same meeting, Mike Turner presented work that he and Lawrence Krauss had been developing. What if the *total* Ω is one, but the dark matter density is just what it appears to be, $\Omega_m = 0.3$. These statements could both be true if something besides dark matter contributes to the energy density of the universe. What if the rest of the energy density is made up of smoothly distributed dark energy, so that Ω_Λ, the energy density associated with the cosmological constant, is a significant fraction of the universe? A similar set of arguments had been advanced by Paul Steinhardt and Jerry Ostriker in a recent *Nature* article. When I want to tease Jerry (always), I say that he applied deep ideas, noting that if $\Omega = \Omega_m + \Omega_\Lambda$ and if you just know in your heart that inflation means $\Omega = 1$ and you know from observation that $\Omega_m = 0.3$, then using the powerful theoretical method of "subtraction" even I could compute that Ω_Λ = 0.7. If the universe isn't made of dark matter, it must be made of dark energy. Theory really isn't so difficult.

Like all effective teasing, this is a little unfair, because there is another cosmological fact that Turner and Ostriker and Steinhardt could match with Λ, but could not match without it. That is the age of the universe. If the Hubble constant was something like 80 kilometers per second per megaparsec, as initial observations with the Hubble Space Telescope then suggested, there was a real problem. If Ω_m = 1, then the true age is two-thirds of the apparent age, because of deceleration. The apparent age, the Hubble time, is 12 billion years, and two-thirds of 12 is 8 billion years. This was not in good accord with the best measurements for the ages of the globular cluster stars, which appeared to be older: at that time, the experts put the globular cluster ages around 15 billion years. So, according to the logic of the case, there was a problem with Ω_m = 1, and a need for Ω_Λ to make up the balance of the mass–energy in the universe.

Turner, Ostriker, and Steinhardt are excellent debaters. They make a case like prosecuting attorneys. Listening to the presentation, you are inexorably swept along to the conclusion. Except science is not law. Convincing the jury is not enough. Although you would always like to convince the jury of informed opinion that your view is correct, the data have the final word. And Saul Perlmutter had presented data on supernovae that indicated deceleration and Ω_m near one, and that left no room for this reincarnation of the cosmological constant. Theorists are valuable as long as they are stimulating. It is not so important for them to be correct. Observations, on the other hand, are useful only when they are right.

The result that Saul presented at Princeton was published in July 1997 in *The Astrophysical Journal*. Their best estimate of Ω_m was 0.88 and they asserted that the data put the strongest known upper limit on the energy density associated with the cosmological constant of Ω_Λ less than 0.1. Of course, this was just a preliminary result, and the SCP promised much more data in the coming year, but they had put their stamp on the field. They took the theoretical argument head on, saying their results were "inconsistent with

Λ-dominated cosmologies that have been proposed to reconcile the ages of globular cluster stars with higher Hubble constant values."[1]

The group at Lawrence Berkeley Lab were cool to the idea of another group working in the same area. But our own high-z team didn't need their permission. We just had to make the case to the people who decide about scarce telescope time that it was worthwhile to have another team at work on this important subject. I thought we had a good case because of the depth of experience of our team in studying supernovae over the past 20 years and our collective mastery of the tricky problem of doing accurate photometry on faint objects. People on our team had built up the entire sample of nearby supernovae that either team would need to compare to distant supernovae. We had invented the techniques for making SN Ia into good standard candles using colors and light curve shapes to compensate for dust and intrinsic variations among SN Ia. Besides, this was an important problem and it would be good to have two teams work on it to see if the answers agreed.

This argument was successful, we got assigned the telescope time, and we began to search for and observe high-redshift supernovae. Brian Schmidt and Nick Suntzeff catalyzed the formation of the high-z supernova team. Brian was on his way from the Harvard–Smithsonian Center for Astrophysics (CfA) to the Australian National University. Our gang at the CfA included Pete Challis, Peter Garnavich, Saurabh Jha, and me. From Cerro Tololo, Nick engaged Mark Phillips, Mario Hamuy, Bob Schommer, and my former Ph.D. student Chris Smith. At Berkeley, there was Alex Filippenko and Adam Riess, who had finished his Ph.D. at Harvard and was now a prestigious Miller Fellow at Berkeley. Alex (who had himself been a Miller Fellow a decade earlier) had been part of the LBL team, but once we got our high-z act together, he chose to work with us.[2] We were very glad to have him. Berkeley graduate students Alison Coil and Ryan Chornock pitched in later. We had strong connections with the European Southern Observatory, with Bruno Leibundgut, who had been my postdoc, and Jason Spyromillio, who had done beautiful work on SN 1987A. We enlisted help from the University of Washington, too, with Craig Hogan, Chris Stubbs, and his students Alan Diercks, David Reiss, and Gajus Miknaitis. Alejandro

Figure 10.1. **The high-z team**. A large fraction of the high-z team in a single place for 1/30th of a second in the summer of 2001. Courtesy of Robert Kirshner, Harvard-Smithsonian Center for Astrophysics.

Clocchiatti moved from Texas to Chile, which gave us a shrewd spectroscopist in Santiago to help press the work ahead.

As time went by, some students finished their degrees and left, while new ones joined. And, as the methods and aims of the program evolved, we added Ron Gilliland at the Space Telescope Science Institute, and John Tonry at the University of Hawaii and his student Brian Barris. Our greatest technical achievement was to make up a team cover sheet for talks and proposals, with all the logos of the institutions involved. But it betrayed our prejudice. The cover sheet said, "The High-Z SN Search" and went on to say "Measuring Cosmic Deceleration . . . with Type Ia Supernovae." In the end, we did *not* measure cosmic deceleration, but something else.

By astronomical standards, where a typical research group has a faculty member, perhaps a postdoc, a student or two, and a pet dog, this was a big group. On the other hand, compared to particle physics research teams of the type they were accustomed to assembling at LBL, this was an intimate club. Our team needed to be big because of the peculiar requirements of a supernova search. New supernovae are like fresh fish. If you don't use them right away, they spoil. So our search and follow-up had to be carefully orchestrated and intense. Just as for Zwicky at Palomar in the 1930s, the Vikings before us, or the Calán/Tololo search, the rhythm of the observing was set by the phase of the moon. First you need a template—the "before" image taken in the dark phase of the moon. You wait a month for the moon to cycle through its phases, then repeat the same field in the next dark run.

Now the clock is running. There may be new supernovae in your data, and you have to find them before they fade into uselessness. Working round the clock, fueled by Chilean pizza, Tucson tacos, or Kona coffee, team members struggle with the software to get all the images aligned, blurred to match, scaled to the same brightness levels, and subtracted. Sometimes it goes smoothly, sometimes not. But always there is a sense of urgency.

The automated software spits out postage-stamp-sized images of possible candidates: places where there is a 5σ something on the second image that wasn't on the first. Not everything that glitters is gold. Somebody has to look at every one of these events to see if the software has done something stupid. There are satellites, asteroids, electronic noise, diffraction spikes, bad subtractions, bad columns, hot spots, cosmic rays. And supernovae. Somebody has look at the image to tell the difference. It is tedious, hard work done under pressure. The clock is ticking, not just because the supernova might be fading, but because the follow-up observations are already scheduled and people are moving into position to take those data. But they can't take data if we don't find the supernovae.

For a typical observing run, we take dozens of images with the largest CCD cameras we can get our hands on. The big cameras have 100 million pixels—about 30 times the size of a "high-resolution" digital camera you can buy at Circuit City. The data from a single exposure fill 30 good-sized monitor screens, and a typical night produces 30 images. So that means we need to scan through almost 1000 screens-full. Each image has thousands of galaxies of about the right distance to be interesting sources of supernovae for cosmology. So if there's a supernova every 100 years in a typical

Figure 10.2. **Suprime: a giant CCD camera**. The advent of very large electronic cameras is the technical advance that made high-z supernova searches practical. These cameras have close to 100 percent efficiency using silicon charge coupled devices (CCDs). This one has about 100 million pixels compared to 3 million in a high-end digital camera you can buy today. Courtesy of Subaru Observatory.

galaxy, we should see several in each observing run. If the weather is good. If the software works properly.

While some team members are sifting the data for new stars, others are already on the way to big telescopes to follow up the discoveries. It is the strangest form of observing. Usually, you do meticulous preparation long in advance. You make a list of your targets, make finding charts of their locations so you can identify them at the telescope, and think through just how to use your nights so you don't waste observing time. But for the supernova follow-up, there's no way to do all this in advance. So you travel to Tucson or to Kona or to La Serena with nothing prepared. While you are in the airplane, teammates are, you hope, generating a list of good candidates: new dots on the images that might be supernovae half-way across the universe. It's a heart-wrenching way to observe.

While we try to provide a few days between the search and the follow-up, sometimes that margin gets eaten up by glitches in the data processing. Then the sickening possibility of wasted time on the largest telescope in the world begins to gnaw at the observers. Alex Filippenko could be at the Keck Observatory in Hawaii waiting with the suppressed tension of a drag racer at a red light while Pete Challis is still slaving away in Chile, sorting reality from illusion.

On a calm day, Alex is a bundle of nervous energy. This relentless attention has served him well—Alex has become one of the most productive astronomers in the world. Slender, intense, and focused, he has the fast-twitch muscles of the star tennis player he is and the eating habits of a fast-food junkie. On the afternoon of an observing run, Alex snarfs Cheese Doodles while his bouncing leg communicates his anxiety. Has Pete put the targets at the team website?

"Not yet."[3]

When twilight begins in Hawaii, Alex walks across the Keck parking lot to the nearby McDonald's and buys a bag of Big Macs. If the targets are still not posted, the tension is contagious. Alex becomes like Sherlock Holmes without a case. In *The Adventure of the Wisteria Lodge*, Sherlock says, "My mind is like a racing engine, tearing itself to pieces because it is not connected up to the work for which it was built." But once Brian Schmidt and Peter Garnavich get the observing list in order, the Keck dome is open, and it's time to get to work, Alex is the best guy to have in the pilot's seat because he focuses all that energy on the task at hand. Paying attention doesn't make the photons come in faster, but it helps you anticipate what to do next, and avoid wasting precious telescope time. Later in the night, Alex refuels with hamburgers, without regard for temperature, freshness, or the texture of the congealed cheese, and washes them down with strawberry soda. While others' attention drifts, Alex never flags, squeezing every minute of data from a night at the mighty Keck.

We get spectra of the supernova candidates at the Keck or the Very Large Telescope that ESO runs in the north of Chile. A new dot might be a supernova, but it might be something else. A spectrum will tell you if you've selected a variable quasar (oops!), a type II supernova (close but no cigar), or the SN Ia we know how to mold into the best of standard candles. The spectrum will also reveal the redshift, so we know where to put the supernova on one axis of the Hubble diagram.

But this is hard work. The supernova light is only about 1% of the light coming into the spectrometer from the sky. So it requires meticulous subtraction to see clearly what you've got. And you need to make decisions rapidly, to work through the list of candidates to find the genuine SN Ia. This combination of careful work and rapid decisions is a volatile mix. It's best to divide the labor, with somebody who is computer-nimble (under 30) doing the data reduction, a skilled operator who knows the telescope and the instruments at the controls, and someone in the role of Mr. Spock to provide logical advice on what to do next. Add in uncertain weather, balky instruments, and jet lag to brew a cauldron of stress.

But the results have been very good. Even with mediocre weather, we usually find several type Ia supernovae per search night in the redshift range from 0.3 to 0.8 where the effect from cosmology is most accessible. For example, in 1999, two nights of searching at the Canada–France–Hawaii Telescope (CFHT) in Hawaii and at the Blanco Telescope at Cerro Tololo provided a list of 20 objects with spectra, 12 of which we were confident were SN Ia, which ranged in redshift from 0.28 to 1.2. This is the deep water where you can learn the history of cosmic expansion.

Then we measure the light curve. We need to know how bright the supernova was at maximum light. And we also need to measure the shape of the light curve to determine whether we are dealing with a typical SN Ia, one that was a little extra bright, or one that was a bit of a dim bulb. Plus, to measure the effects of dust absorption, we measure the supernovae through more than one filter to get the color. Most of the information about the shape of the light

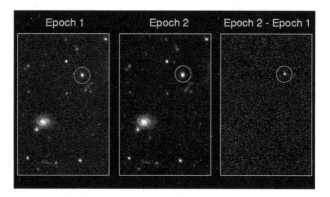

Figure 10.3. **Subtracting images to find supernovae.** The image from a month ago is subtracted from last night's image to reveal a new supernova. The image area shown is about 1/1000 of the full area provided by the CCD camera. Rapid processing of dozens of image pairs demands nimble software and a dedicated team of searchers. Courtesy of Brian Schmidt, Australian National University.

curve comes in the first month after maximum, so that's where we concentrate our effort.

But cosmological expansion not only shifts spectrum lines to the red, it also slows the ticking of distant clocks, so the radioactive decay that powers distant supernovae appears to run more slowly at high redshift, making our follow-up job a little less urgent.[4] For a redshift of one, 30 days elapsed at the supernova corresponds to 60 days for the observer at the telescope. So to find out how fast the supernova declined in the first month after maximum (with time measured at the supernova), you need to observe it several times in the next *two* months here on Earth.

Even then you are not done. Supernovae are bright objects, but they are not ordinarily as bright as the galaxies of 100 billion stars in which they erupt. So, even if you are very careful, the host galaxy can add a significant amount of galaxy light to the supernova light you want to measure. You need to subtract the galaxy light. We wait for a year, then come back and take a really good "after" image that will show the galaxy, which, like the Cheshire Cat's smile, is still present after the supernova has faded. This makes the whole process quite sluggish. A supernova that you discover in 1995 needs to be revisited in 1996, and it wouldn't be surprising if it took well into 1997 to pass the strict photometric quality control of Nick Suntzeff. So even though we were working diligently starting in 1995, we didn't have much to say until the end of 1997.

All our discoveries, all the spectra, and most of the light-curve data came from the ground. The Hubble Space Telescope is the most wonderful telescope of our time, but it makes its super-sharp images of only a small patch of the sky. It is an effective tool to search for supernovae only when you are interested in extremely distant galaxies that are cheek-by-jowl in a deep Hubble field. To search wide areas to a moderate depth, we have used 4-meter telescopes with big CCD cameras at Cerro Tololo and the CFHT on Mauna Kea.

HST does make beautiful images. It is above the blurring effects of the Earth's atmosphere and it is (now) a nearly perfect optical system. This can help solve the problem of measuring light from a supernova that is on top of a galaxy. The angle between the supernova and the galaxy is often less than 1 arcsecond—about the amount that the Earth's atmosphere blurs the image for both the supernova and the galaxy. We subtract the galaxy light measured later, but the results are never perfect. We do better by using the Hubble Space Telescope to take a series of pictures of the fading supernova. In each of them, the image of the supernova is a small hard dot, only 1/100 the area of a ground-based star image, and the separation of galaxy light from supernova light is much more precise. Precision matters because we are using the apparent brightness of the supernova to measure the history of cosmic expansion, and the expected effects are small.

But there is a price for using the Hubble Space Telescope. Bureaucracy. The paperwork associated with HST observing is somewhere on the scale of personal inconvenience between doing your tax return and enduring a root canal. The Space Telescope is in a low orbit, circling Earth every 90 minutes or so. It operates as a robot—ground control loads a long list of instructions into an onboard computer every week, and then HST plods down its to-do list, moving to the objects of interest, locking on to guide stars, acquiring data with the cameras or spectrographs, and then sending

SN 1997cj

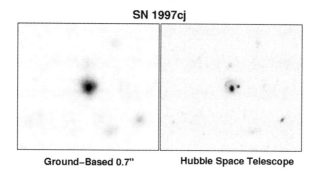

Ground–Based 0.7"	**Hubble Space Telescope**

Figure 10.4. **HST and Ground-based images of SN 1997cj.** Sharp images from the Hubble Space Telescope make accurate measurements of supernovae much easier. Courtesy of Peter Garnavich; University of Notre Dame/NASA.

the bits to the ground by radio. Since this intricate dance is taking place without human intervention, the crew at the Space Telescope Science Institute in Baltimore likes to get everything set and checked well in advance. They don't like surprises, and they don't like last-minute changes to that list of instructions. For some reason, they believe the safety of the telescope is more important than leaping immediately to implement our desires.

So their rule is: tell us where you want to observe a month in advance. Now, for ordinary observing with HST, this is a reasonable rule. It gives the scheduling wizards at the Institute time to build an efficient schedule and to check and double-check the telescope's instructions before they are sent up. Next week's instructions are reviewed during a work week in Baltimore, and the "SMS Load" goes up to HST through NASA's communications network on the weekend.

But what if you want to observe a supernova? How are you going to schedule *that* a month in advance? We have some experience with that. I am the principal investigator for a program we call SINS (Supernova INtensive Study), whose aim is to use HST to learn more about nearby, bright supernovae by using HST's unique ability to observe in the ultraviolet. We've had good success in working with the Space Telescope Science Institute to get new objects into the schedule on short notice, including the nearby SN Ia SN 1992A.

But these "Target of Opportunity" observations have the disruptive quality of a 2 A.M. fire alarm in a college dorm and the Institute, quite rightly, doesn't want too many of them because they demand so much staff attention.

As chief SINner, I have learned over the years that the natural rhythm of the Institute responds best to supernovae reported to them on a Tuesday. Then they can revise the instruction set during the balance of the week, send it up on the weekend and, if you are very lucky, get your target observed as soon as the next Monday. Elapsed time: as little as six days. They don't respond so well to a supernova found on a Thursday night. If you call on Friday morning, they'll say "too late" to change the instructions for next week. At best, they might put you on the docket for a week from next Monday and it might be a week from Sunday. Elapsed time: 11 to 18 days. Does this procedure challenge the limits of human intelligence? Not really. But it *is* rocket science.

The Space Telescope can't point just anywhere on the sky—it has to avoid the sun, the moon, and especially the Earth, all of which are changing position as seen from a low Earth orbit. So, the problem is to find supernovae at places on the sky and times specified a month ahead, and to report those to the Institute early in the week. This is not quite as crazy as it sounds. Luckily for us, Pete Challis at the CfA used his understanding of the inner workings of the Space Telescope system to puzzle out how to do this. We enlisted Ron Gilliland at the Space Telescope Science Institute to help us solve this problem. Of all the people who use HST, Ron is the most successful in thinking through exactly how the instruments and operations of this complicated machine can be used to do unusual and valuable science.

Here's what we do. The telescope time we get assigned at CTIO or in Hawaii for supernova searches comes through the usual mechanism of Time Allocation Committees and telescope scheduling for big ground-based telescopes. That's done six months in advance. So we know *when* we are going to find distant supernovae. If we find any at all, we are going to find them right after the nights when we observe. If we have a choice, we try to make the discoveries on a Saturday so we have time to sort things out before we choose our

HST targets. And we know *where* we are going to find them. We are going to find them in the fields we observed in last month's dark run and for which we are going to make observations this month. We know where our target fields are. So we know we are going to find supernovae *when* we have observing time and *where* we point the telescope.

The search fields are about half a degree across, which is precise enough for the Space Telescope schedule to be constructed. This needs to take into account a zillion technical details. Ones I know of are the timing of the telescope's orbit, the direction to the sun, and the time it takes for the telescope to turn from the previous object to ours, moving at the stately pace of the minute hand of a clock. And there are lots more I don't have the neurons to know (but Ron Gilliland does). We tell the Institute boffins the precise position of the newly discovered supernovae we want to observe, they insert those details into the schedule, check, double-check, transmit, and execute. Elapsed time: about a week.

Is it worth all this bother? Absolutely. In 1997, the high-z team pulled this off, discovering supernovae on schedule at CFHT or Cerro Tololo, getting their spectra at Keck and at the MMT in Arizona, early light curves from the University of Hawaii's 88-inch telescope, and after delivering the precise target list on a weekday, we obtained a beautiful sequence of observations with HST starting one week after the Keck spectra and extending over the next 80 days.[5] While our original motivation for using HST was the wonderful imaging that makes photometry more precise, we also benefited from the absence of weather and the fact that moonlight doesn't light up the sky when you are above the atmosphere. The observations took place exactly as planned, which hardly ever happens on the ground, and we could time them in the optimum way to learn about the light-curve shape.

One difficult part of these measurements was making certain that the measurements from HST and from the ground agreed. To do this, we carefully matched ground-based and HST measurements of 15 background stars in the HST images that did not vary, which were bright enough to see from the ground, but not so bright they overwhelmed the HST's CCD detector.

Circular No. 6819

Central Bureau for Astronomical Telegrams
INTERNATIONAL ASTRONOMICAL UNION

Mailstop 18, Smithsonian Astrophysical Observatory, Cambridge, MA 02138, U.S.A.
IAUSUBS@CFA.HARVARD.EDU or FAX 617-495-7231 (subscriptions)
BMARSDEN@CFA.HARVARD.EDU or DGREEN@CFA.HARVARD.EDU (science)
URL http://cfa-www.harvard.edu/iau/cbat.html
Phone 617-495-7244/7440/7444 (for emergency use only)

SUPERNOVAE

P. Garnavich, Harvard-Smithsonian Center for Astrophysics (CfA), reports that the High-Z Supernova Search Team (*IAUC* 6160, 6646) has discovered nine supernovae on CCD images taken with the Cerro Tololo Interamerican Observatory (CTIO) 4-m telescope by R. Schommer (CTIO), B. Schmidt (Mount Stromlo and Siding Springs Observatory), and S. Jha and P. Challis (CfA). The supernovae were identified by subtracting images taken on 1997 Dec. 29–30 UT from those obtained on 1998 Jan. 23. The candidates were confirmed with spectra and images taken on Jan. 25–26 and Feb. 1 with the Keck-2 telescope by A. V. Filippenko, A. G. Riess, and D. C. Leonard (University of California, Berkeley).

SN	1998 UT	α_{2000}	δ_{2000}	I	z	type
1998F	Jan. 23	$4^h16^m50^s.13$	$-5°44'59''.6$	24.5	0.52	?
1998G	Jan. 23	8 03 37.02	$+6$ 10 13.9	22.8	0.30	II?
1998H	Jan. 23	8 04 51.47	$+5$ 36 39.3	23	0.66	?
1998I	Jan. 23	8 04 51.56	$+5$ 15 47.7	23.6	0.89	Ia
1998J	Jan. 23	9 31 10.48	-4 45 36.5	22.5	0.83	Ia
1998K	Jan. 24	4 13 42.86	-5 50 45.2	23.8	?	?
1998L	Jan. 24	11 33 36.63	$+4$ 35 04.6	23.6	?	II?
1998M	Jan. 24	11 33 44.37	$+4$ 05 13.4	23.3	0.63	Ia
1998N	Jan. 24	11 33 29.39	$+3$ 51 12.5	23.1	0.26	?

Each supernova is within $2''$ of its host galaxy's center. A foreground galaxy with $z = 0.02$ is centered $2''.5$ southeast of SN 1998M. Finder charts may be requested by sending e-mail to pgarnavich@cfa.harvard.edu.

4U 1608–52

W. Cui, Center for Space Research, Massachusetts Institute of Technology; and J. Swank, Goddard Space Flight Center, report on behalf of the ASM team at MIT and RXTE Science Operation Facility: "The real-time ASM lightcurve indicates that 4U 1608–52, a recurrent soft x-ray transient source, started an x-ray outburst on Jan. 29. Rough daily-averaged 1.3–12-keV fluxes: Jan. 29, 25 mCrab; 31, 21; Feb. 1, 28; 2, 121; 3, 261; 4, 491. A pointed observation was carried out with the PCA and HEXTE detectors aboard RXTE on Feb. 3.82 UT, and the measured flux was consistent with the ASM results. More pointed RXTE observations have been planned. Observations at other wavelengths are encouraged."

1998 February 5   © Copyright 1998 CBAT   *Daniel W. E. Green*

Figure 10.5. An International Astronomical Union Circular from the Bureau for Astronomical Telegrams reporting the results of two nights of searching in 1998. Some of these supernovae were observed with the Hubble Space Telescope. Courtesy of the Central Bureau for Astronomical Telegrams.

Peter Garnavich, a postdoc working with me at the CfA, now on the faculty at Notre Dame, took responsibility for getting the HST data reduced, and by the end of 1997, we were finally in a good position to say our first words about cosmology. Based on the data in hand, we did not agree with the LBL team's earlier conclusion as discussed in Princeton and published in July. They had found evidence for deceleration, corresponding to Ω_m near 1. In their data, that meant the supernovae appeared a little brighter than

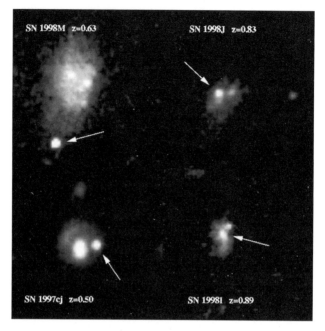

SN 1998M z=0.63   SN 1998J z=0.83

SN 1997cj z=0.50   SN 1998I z=0.89

Figure 10.6. **High-z supernovae observed with the Hubble Space Telescope.**
Courtesy of Peter Challis; High-z team/NASA.

they would in a freely coasting lightweight universe. When Garnavich plotted up the data, our supernovae showed no such effect. Although the data were too scanty to tell us the whole history of cosmic expansion, they were adequate to rule out $\Omega_m = 1$. We worried a little that the LBL team had published a contrary result. But this was hard work, and there were many ways to go wrong. We decided not to worry too much about the other guys, to judge our own measurements by our own internal standards, and to hope for the best.

The vivid way to state Garnavich's conclusion is that we showed that the universe would expand forever. That seemed like interesting news, so we sent in an abstract for the forthcoming meeting of the American Astronomical Society (AAS), which was going to be in Washington, D.C., in January 1998. We didn't yet have enough data to say whether there was or was not cosmic acceleration, so we were silent on that point.

Meanwhile, the other team was changing its tune. In July 1997, they had published an article in *The Astrophysical Journal* with data pointing to large Ω_m. Now, at the end of 1997, we heard they had a new result submitted to *Nature*, with an HST observation of their own, which claimed the opposite. With the addition of just one new supernova, augmenting their sample of seven, they now found that their evidence pointed the other way, toward low Ω_m. The one new supernova had observations from HST, so it was presumably better data and, if calibrated carefully, carried more weight than the earlier work. Still, for one object to turn July's conclusion on its head seemed extraordinary. We had no way to check their work, since neither of their papers published the details of the light curves and spectra. In any case, the SCP also submitted an abstract for the upcoming AAS meeting (which we read carefully!) stating clearly that they now found evidence for low Ω_m. On the subject of cosmic acceleration, though, that abstract was silent.

In the Fall of 1997, the Institute for Theoretical Physics (ITP) at the University of California, Santa Barbara, sponsored by the National Science Foundation, held a workshop on supernovae. I had never taken a sabbatical in 21 years as a university professor, my personal life was in transition, and this seemed like the right time for a break from the routine. Unlike New England, where a nice day is a rare thing highly prized, Santa Barbara is a place where almost every day is pleasant. People lose their sense of urgency. Play tennis? Oh, maybe tomorrow. It will be nice tomorrow. Physicists are not entirely immune to the charms of this place, but they run on more tightly wound internal springs than most Santa Barbara residents. Play tennis? Oh, maybe tomorrow. Today let's figure out supernova light curves.

Although the ITP is really a place for theoretical physics, and it would be false to say I am a theorist and misleading to say I am a physicist, they treated me very well. Sort of like a pet Bernese mountain dog. A little out of place in Santa Barbara, not very good at retrieving ducks, but amusing. As a service for the ITP, I gave a public talk for the local community on high-redshift supernovae and the quest for understanding cosmology. Unfortunately, in the fall of 1997, we were not quite to the moment of having an important result—we knew how to do the problem, we had some data in hand, but we didn't quite have the answer.[6]

As a reward for my public-spirited behavior, David Gross, the director of the Institute for Theoretical Physics, and Adam Burrows, an organizer of the supernova workshop, gave me a made-up union card in the International Brotherhood of Theorists. Decorated with a spilled coffee cup and stubbed out cigarette butts, it declares the theorist's self-referential motto: *Cogito ergo sum*. I carry it around in case I think I am a theorist.

The real theoretical physicists at the ITP were very attentive to cosmology—it is a fast-moving field where the data might demand new physics. The cosmological constant was a well-known problem in theoretical physics. My office had a spectacular view of the ocean, including surfing undergraduates and swimming dolphins, but right across the hall at the ITP was Sean Carroll, a young postdoc who had been one of the brightest and most interesting astronomy graduate students at Harvard (as a student, he shared an office with Brian Schmidt).[7] Sean was a precocious author of a review article on the cosmological constant written in 1992, with Bill Press of Harvard and Ed Turner of Princeton. The review summarized the problem from the point of view of astronomers, looking for evidence, and from the point of view of theoretical physics, reasoning from the nature of particles and fields. Though the value of the cosmological constant allowed by astronomical observations in 1992 might have been as large as Ω_Λ equals 1, the simplest theoretical prediction gave $\Lambda = 10^{120}$ (that's 1 followed by 120 zeroes!). More sophisticated theoretical reasoning could make this 10^{80}, or perhaps 10^{50}, but there was no theoretical reason that very bright people could think of why Λ should be a small number like 0.1 or 0.6 or even 17. Faced with the astronomical reality of a small (compared to 1 with 50 zeroes) cosmological constant, many theorists suspected it would be exactly zero. This is a good second guess. But not everything that's infinite cancels out.[8]

Sean Carroll's article made it clear that there was no positive evidence for a value of Λ that was different from zero, just upper limits from the absence of various effects that Λ would cause. Looking backward, it is amusing to see that Sean's 1992 article makes no mention of supernovae as a possible way to see if the universe was accelerating, as a small Λ might make it do. The work of the Vikings at the Danish telescope in Chile in 1985, which was aimed at this goal, simply hadn't made it onto the theoretical radar screen. But as a problem in theoretical physics, the cosmological constant was a real riddle.[9] Steven Weinberg, a distinguished particle physics theorist, has called the cosmological constant "a bone in our throat."

Even though I didn't yet have anything definitive to report, Sean's antennae were up for any hints that the cosmological constant might become respectable again. The ITP is a center for the revolution in particle physics that is trying to build a new theory for the quantum mechanical forces that operate at the subatomic level *and* that incorporates gravity. General relativity had been around since 1916, and quantum mechanics was developed in the 1920s, but there was still no quantum theory of gravity that united these two powerful pillars of twentieth-century physics, and building that bridge was a serious quest for theoretical physics. Right down the hall from me were people working on developing string theory, which holds out the best hope for making a single theory that covers *all* the known forces. One challenge for this new theory is to provide a natural explanation for a small value of the cosmological constant by connecting the quantum world with gravity. You really didn't need an astrophysical measurement of Λ to know it was small compared to 10^{120}, so for years the subject was mostly a private conversation among the theorists. This was about to change.

For the supernova tribe, the "work" of the workshop included discussing the physical origin of the effect we were using to make type Ia supernovae better standard candles. What accounted for the fact that some of the thermonuclear supernovae were extra bright, and some were dim? And why were the light curves different for bright supernovae and dim ones? Those seemed like tractable questions, and the assembled explosive types, including Friedel Thielemann (recently on the Harvard faculty, now Herr Professor Doktor in Basel), Adam Burrows, Ken Nomoto, Wolfgang Hillebrandt, my nocturnal tag-team wrestling partner Craig Wheeler, Dave Arnett, past CfA postdoc Phil Pinto, and my one-time student Ron Eastman, seemed like people who could help answer them.

After all, it was not enough to have a practical, empirical way to use SN Ia to measure precise cosmic distances. If you didn't also understand them, you might get fooled when you looked at distant galaxies where the chemistry was different and the stars were, on average, younger. One possibility was that both the bright and the dim supernova came from very similar objects crammed up against Chandrasekhar's upper mass limit for white dwarfs, but that some had more radioactive power for their light curves because they fused more nickel in the explosive flame that ripped through these stars. An alternative was that some of the exploding white dwarfs were not at the Chandrasekhar limit, but came from lower mass stars that exploded in a different way that accounted for the range in SN Ia brightness.

The decline rate seemed to have something to do with the atmospheres. If there was a lot of heat supplied from radioactive nickel, the atmosphere might stay warm and opaque longer, making a slower decline rate for the intrinsically bright objects. The dim ones would cool off and turn transparent sooner. These were just ideas, and they needed to be worked out in more detail to become convincing explanations for the data. Santa Barbara was a place to do that work. We could always play tennis tomorrow.

As Thanksgiving approached, the air in Santa Barbara was full of talk about exploding white dwarfs of differing light output when Gerson Goldhaber, a senior member of the Supernova Cosmology Project, came to tell us what they were doing. Gerson comes from a distinguished family of physicists: husbands and wives, uncles and aunts, cousin and nephew physicists from coast to coast. Gerson was a veteran of experimental particle physics, having been in the middle of work on exciting new particles of the 1970s that led to the physicists' Standard Model. Well known and highly respected by the physicists, Gerson was in slightly unfamiliar terrain among the astronomers.

An imposing gray-bearded figure, Gerson spoke slowly in a rich European accent, pulling gently on a pair of broad suspenders that stretched over his convex figure. Like many other successful physicists, he had succumbed to late-onset astrophysics, taking on the challenge of searching for high-redshift supernovae with the same intensity he used to find the charmed mesons. At LBL, they had spent several man-years building their own computer software to sift the supernovae from repeated images, while our high-z team had woven together software from existing astronomical programs that did equivalent tasks. Astronomers and physicists are tribes from different parts of the forest, and Gerson didn't know many of the people in the room or that the hot topic among the supernova theorists at Santa Barbara was to account for the differences in light output of type Ia supernovae.

He started his talk with a picture of a candelabra and spoke of standard candles. He told us that supernova explosions were all identical. By measuring the apparent brightness, the LBL group had developed a method to measure the distances to supernovae and to measure the history of cosmic expansion. I thought it was useful, if not polite, to break in.

"Gerson. The people around this table are trying to understand the reason why type Ia supernovae are *not* alike. It's too simple to say, at least to this group, that all SN Ia are identical."

Gerson didn't like it one bit. He bristled, then turned formally to Friedel Thielemann. "Mr. Chairman, must I endure these interruptions?"

Friedel smiled and said this was a workshop, that the interchange of ideas was important, and that a free discussion was our style. Then he gave me a glance that meant, "Bob, shut up, and stop causing trouble."

At dinner that night at a French restaurant in downtown Santa Barbara, I was polite, if not useful. Gerson's afternoon talk had been mostly about methods and didn't have much about the LBL team's latest results. I was interested in learning exactly what had made them change their conclusions by 180° from July to November. But I wasn't able to learn anything about new results at Berkeley from Gerson. He was very discreet, and did not discuss the *Nature* paper that was being refereed (but not by me!). Gerson deftly steered the conversation to the comparative merits of French restaurants near CERN, the giant particle accelerator near Geneva. My fiancée, Jayne

Loader, had no trouble drawing him out on this delicious topic. Gerson didn't seem at all interested in the progress of our high-z team. I modestly volunteered we were several months behind them.

"You mean several years," Gerson said.

I didn't say anything, but toward the end of 1997 we were already beginning to see hints of something more interesting than just a low-Ω_m universe that would expand forever. Adam Riess was assembling our high-z data at his office in Campbell Hall, on Berkeley's main campus, just down the hill from the LBL team. Adam thought he was beginning to see evidence for cosmic acceleration. Our data showed that the distant supernovae were *fainter* than they would be in a low-density universe. Faint supernovae meant larger distances. Larger distances meant cosmic acceleration. Every time he tried to use the data to determine Ω_m without Λ the value for the mass kept coming out *negative*. That wasn't right. So he added in Ω_Λ, and the best fit to the data points kept giving a value of the cosmological constant that was bigger than zero. As the data trickled in, Adam added more supernovae to the analysis. The statistics were beginning to make the case for the cosmological constant.

I did not like this result. The cosmological constant was a bad companion. For the past 50 years, every sensible paper either began with "we assume $\Lambda = 0$," or just assumed it without saying so. Even if Jerry Ostriker and Paul Steinhardt were making the case for Λ, and Mike Turner at Chicago had tried out Λ in recent years, they were just theorists being provocative. This was not a Greek letter that a well-behaved observer ought to be seen with. How could we be sure there wasn't a dumb mistake somewhere in the long chain of data reduction? Had somebody else checked the numbers?

Adam said that Brian Schmidt concurred with the analysis. I still thought this was a result that would go away as we accumulated more data, and I did not like the idea of going out on a limb and then being forced to crawl back. I had done that once with SN 1987A.

Summoning my dignity, I said, "Adam, the punishment for being wrong should be as big as the reward for being first."

"Reward?" Adam said. "You're going to give me a reward?"

In December 1997, Jayne, our bull terrier Albert, and I decamped from Santa Barbara for a few weeks in Pasadena at Caltech. Fritz Zwicky was long gone, my thesis advisor, Bev Oke, had retired to Victoria, B.C., Jim Gunn had been in Princeton for 17 years. Leonard Searle had retired. Wal Sargent was still there, but the Robinson Lab was a different place. In a way, the mid-1970s had been a high-water mark at Caltech. When the 200-inch reigned supreme, Palomar Power dominated the astronomical scene. Then followed an unpleasant two decades of parity as 4-meter telescopes sprung up around the world in the 1970s and 1980s to challenge the hegemony of the Big Eye. This was good for me, good for science, but not so great for Caltech.

Now, the Caltech astronomers once again had the kind of advantage they liked. With Caltech holding one-third of the time on the two Keck telescopes, a Caltech astronomy professor once again had about 10 times the observing power of anyone else. That's the way they like it.

They set me up on the second floor in Robinson Lab, the quarterdeck where most of the faculty had their offices. This was rarefied air for someone who had worked in the engine room of the second sub-basement. I couldn't even find my way down to 0013. The way was blocked with radio astronomers. I shared the second-floor office with Richard Ellis, who was visiting from Cambridge, where he was Plumian Professor, Eddington's successor. Richard had been leading the way in studying how galaxies evolved over time, and had also contributed to studying high-redshift supernovae. Richard had worked with the Danes in the subject's pre-history to follow up their supernovae, and Richard was now working with Saul Perlmutter, helping the LBL team with observations at the Isaac Newton Telescope and elsewhere.

One December day at the end of 1997, Richard and I were both in the office while I was having a long telephone conversation about the high-z results with Adam Riess. Miss Manners requires the accidental eavesdropper to act as if one has heard nothing. And Richard was working with the LBL team, so I tried not to give him too difficult a test of his discretion. To Adam I said, "Un hunh," "I see," and "How do you feel about that?" like a psychologist on TV. But the office was too cozy and his brain too active: Richard couldn't help filling in the blanks, and in the end, he could not resist a comment. As he was walking out the door, he turned to me, gurneyed up his Welsh face, and said,

"It *can't* be the cosmological constant."

"It can't be," I agreed, making a face of equally authentic disgust.

The weeks passed quickly in Pasadena while Adam and I went back and forth about the latest results. Did we really believe we were seeing the effects of a cosmological constant? We hadn't reached a resolution by 1 January 1998. Down Colorado Boulevard at the Rose Bowl, Michigan beat Washington State 21–16 and was dubbed the national champion. Go Blue! My son Matthew, a University of Michigan senior, came to town for the festivities. We didn't see all that much of him: Wolverines are everywhere, and Matthew had plenty of friends in southern California to share the triumph.

At the equivalent astronomical event the next week, Peter Garnavich presented our team's evidence on eternal cosmic expansion at the American Astronomical Society meeting in Washington, D.C. Our handful of supernovae favored a low value of Ω_m. Or, more vividly, no slowing down, expansion forever! Go Blue!

Peter shared the podium at a press briefing with Saul Perlmutter. Saul said that they had concluded based on the same seven supernovae from before plus one new one observed with HST that the world was *not* coming to an end. Contrary to their previous result, the SCP now favored a low value for the observed slowing of the universe. Plus, Saul showed an impressive new plot based on observations of 40 supernovae.

What was most interesting was what the SCP did *not* say about their Hubble diagram. At this gathering, with many very interested reporters present, neither team dared to claim they had demonstrated cosmic acceleration, the signature of the cosmological constant. Jim Glanz, then of *Science* magazine, could see where the SCP data might be heading, and wrote a news article for *Science* that tried to anticipate the next step, but at that moment in January

1998, Saul Perlmutter was not ready to say they had seen acceleration. Saul delicately stuck to the subjunctive, as if he were indicating a supposition contrary to fact. Glanz quoted him as saying, "If [the results] hold up, that would introduce important evidence that there is a cosmological constant." Saul wasn't ready to stick his neck out.[10]

Neither were we. Adam Riess had made Peter Garnavich promise not to say anything in Washington about the new data we were working on—he could show the beautiful points from our HST observations, but say nothing about the additional data that was pointing toward Λ.

Peter Garnavich carefully studied the posters the SCP had brought to display at the Astronomical Society meeting. None of them claimed that the SCP had evidence for cosmic acceleration because they had not yet come to a firm conclusion on how to handle "systematic effects," mostly reddening by dust. This was exactly the point I had been trying to make to Saul since that awkward referee report in 1993—if you don't understand the dust, you can't say anything about cosmology.

Now it was time for our team to get serious. The SCP would not sit on the fence indefinitely. They were smart guys and they would either figure out what to say about dust or sweep it under the rug before too long. Was our high-z team ready to climb out on the limb where the data were pushing us? The distant supernovae were coming out about 25 percent dimmer than they would appear in a universe with Ω_m = 0. Dim supernovae implied acceleration, if they weren't dimmed by dust, and our observations in two filters suggested that there wasn't much dust.

How reliable was our result? We had 16 decent objects, 10 with reasonable estimates of the uncertainty from multicolor observations of the light-curve shape to improve the accuracy and precision of the distances. If we believed the formal 3σ error estimates from Gaussian statistics, the chances were 3 in 1000 that this was a bad luck sample in a universe that was actually decelerating. If you believed the error estimates, the odds were about 300 to 1 that we were living in an accelerating universe. Did we believe the error estimates? Did we trust in Gauss?

Well, yes and no.

Yes, the methods using light-curve shapes gave the right-sized errors for samples of nearby supernovae. It was like asking how many Cheerios are in a cereal bowl. You could estimate the number, and also estimate how far off from the true number each sample might be due to chance. Gauss knew how to do this. Bill Press had a recipe for doing this in his mathematical cookbook *Numerical Recipes* and Adam had made it work for the multicolor light-curve shape method.

And no, there might be additional problems in the much sketchier data for distant supernovae that somehow we were not accounting for properly. Maybe we were doing the equivalent of crunching the cardboard with a vigorous twist of the micrometer—getting consistent, but wrong, results.

For the moment, we kept our lips sealed while we tried to decide how seriously to take our own evidence. Bruno Liebundgut had to attend an Alpine conference at the end of January 1998, and bite his tongue. Bruno didn't hurt himself skiing at Moriond, but he had to restrain himself during the discussion of supernova Hubble diagrams. He showed the same data that Garnavich had showed in Washington. In the two weeks that had passed, people had gotten used to living in a universe that would expand forever, and this result now seemed as exciting as cold oatmeal. Bruno did not show the additional data points that made us think we were seeing cosmic acceleration. Somebody from the SCP showed their 42 objects, which looked pretty impressive. But they still did not claim that the data showed we lived in an accelerating universe because they didn't quite know what to do about the "systematics."

Inside our team, we were debating exactly how to proceed—whether to write a quick, short paper that might be wrong but would stake a claim to the discovery of acceleration, or to take more time to write a more thorough paper that would show all the evidence. Everybody on the high-z team weighed in. We had a conference call—always a dubious proposition, but worse when you have participants in Europe and in Australia. Somebody is always half-asleep. We exchanged e-mail. Lots of e-mail.

Adam Riess was doing the heavy lifting for this paper, drawing together all the data, working out the implications, and dealing out the writing assignments. So we all gave him advice. Conflicting advice. After all, this was a collaboration, not an army.

I didn't like the result. I didn't think we were smarter than Einstein and he had tripped on the cosmological constant. I did not want to make a mistake. I hadn't liked being wrong about the progenitor of SN 1987A and I did not want to be wrong about the history of cosmic expansion. On 12 January 1998 (at 10:18:31 A.M.) I wrote,

> I am worried that the first cut looks like you might need some lambda. In your heart, you know this is wrong, though your head tells you [that] you don't care and you're just reporting the observations. . . . It would be very silly to say "we MUST have nonzero lambda" only to retract it next year.

While Peter Garnavich was in Washington, Adam dropped out of sight for a few days to return to New Jersey to marry his MIT classmate, Nancy Schondorf. At the reception, one of Nancy's cousins asked about a news story he had read in the paper that morning. It said the universe would expand forever. Did the groom know anything about this?

"I am familiar with that work," Adam said.

Adam wrote us all a long e-mail (on 12 January 1998 at 6:36:22 P.M.). This was two days after the wedding, just before leaving for their honeymoon, the traditional time for writing e-mail to scientific colleagues.

> The results are very surprising, shocking even. I have avoided telling anyone about them for a few reasons. I wanted to do some cross checks (I have) and I wanted to get a ways into writing the results up before Saul et al. got wind of it. You see, I feel like the tortoise racing the hare. Every day I see the LBL guys running around, but I think if I keep quiet I can sneak up . . . shhhh. . . . The data require a nonzero cosmological constant! Approach these results not with your heart or head but with your eyes. We are observers after all!

Alex Filippenko was all for going ahead fast. His logic was simple. The data pointed toward cosmic acceleration, the LBL team was close to the same conclusion, but not quite ready to take the plunge, so let's publish first. Alex wasn't too worried about being wrong.

"It is possible that there's some sort of subtle systematic effect, but if so, I think it's going to take a long time to figure out."

Writing from Australia, Brian Schmidt was more conflicted:

> It is true that the new SNe say that the complete sample of ~12 objects gives Ω_Λ greater than zero with over 90% confidence. . . but how confident are we in this result? I find it very perplexing and I think we should really try to take the high ground here scientifically. . . . . Let's put out a paper we can be proud of—quickly.

Nick Suntzeff weighed in from Chile with good advice on physical training for Adam.

> I really encourage you to work your butt off on this. Everyone is right. We need to be careful and publish good stuff with enough discussion to make it believable to ourselves. . . . If you are really close to being sure that lambda is not zero—my god, get it out. I mean this seriously—you probably never will have another scientific result that is more exciting come your way in your lifetime.

In the end, we decided to let Gauss be our guide, and to go ahead. If the data said the cosmological constant was a 3σ result, then we were going to say it was a 3σ result and live with the consequences. Less than 1 percent chance of being wrong. Bet $30,000 to win $100. But don't bet your pets.

I had been invited to speak at the Dark Matter meeting that UCLA organizes every other year, but the February dates conflicted with my return to Harvard. So I was driving across America, with Jayne and Albert the bull terrier, seeking motels that take pets, while Alex Filippenko carried the high-z team banner to Marina Del Rey. Gerson Goldhaber and Saul Perlmutter spoke first, showing evidence for time dilation, strong evidence for Ω_m being too small to halt cosmic expansion, and tentative evidence for possible Λ but they were still not quite ready to say that they understood the systematic effects well enough to be certain. Alex presented our team's data and analysis of 16 supernovae at redshifts from $z = 0.16$ to 0.97, comparing them with 27 nearby supernovae from a combined CfA and Calán/Tololo sample. The Hubble diagram for these superno-vae indicated that the universe was not just expanding, and not just destined to expand without limit. Alex said clearly that our superno-vae provided evidence that cosmic expansion *had sped up during the last 5 billion years*.[11]

We were totally unprepared for the press onslaught that started on 27 February. Alex left town to be the tour guide on an eclipse expedition in Aruba. When Adam Riess got to his Berkeley office that day, the phone was ringing. CNN had a camera crew rolling across the Bay Bridge—could they interview him? In 15 minutes? The next day, Adam appeared on *The News Hour*, his father's favor-ite show. The press was really interested in the accelerating uni-verse, but even more interested in how we felt about the results, as if this would somehow affect the universe. Brian Schmidt was quoted as saying, "My own reaction is somewhere between amaze-ment and horror."

Saul Perlmutter's group had been struggling with the same set of questions, doing their best to get it right. Their data pointed toward acceleration, but they weren't quite ready to say they believed that result in January 1998 at the AAS meeting or Moriond or in February at the Dark Matter meeting. They were worried about the right way to treat the absorption of supernova light by dust. We had spent the past five years taking data on nearby supernovae and then working out the way to use light curves and colors to measure dust absorption. We took the plunge in February. In April 1998, Gerson Goldhaber explained his view of this sequence of events to the *New York Times*: "Basically, they have confirmed our results. They only had 14 supernovae and we had 40. But they won the first point in the publicity game."[12]

It was all very well to submit abstracts to meetings, give press briefings about your mental states, and talk at conferences, but the real scientific product is a refereed journal paper. The high-z team concentrated on getting the data into a form suitable for public in-spection with the evidence shown as clearly as possible and the conclusions stated as strongly as the evidence would support. We tried to be our own most caustic critics, probing the weak points of the evidence and exposing the assumptions to debate. By 13 March we had done a job that was not perfect, but good enough. And sometimes good enough is good enough.

We decided to send the manuscript to *The Astronomical Journal* instead of *The Astrophysical Journal* as an inside joke. The other team said they had used a "physics based" approach. Since I didn't know what that meant, it seemed vaguely amusing to use a journal with "astronomical" in its title. Also, we knew the *AJ* publishes things faster. "Observational Evidence from Supernovae for an Accelerating Universe and a Cosmological Constant" was refereed, accepted on 6 May, and appeared in the September 1998 issue. We concluded the abstract of the paper with a long litany of possible sources of error, and then concluded, "Presently, none of these effects appears to reconcile the data with $\Omega_\Lambda = 0$."

All along, we had made the case that it was a good thing for two independent groups to carry through this work. We were very interested to see exactly what the SCP had done. Their paper, "Measurements of Omega and Lambda from 42 High-Redshift Supernovae" was submitted to *The Astrophysical Journal* on 8 September 1998, accepted in December, and appeared in the June 1999 issue.

Although the two programs were independent, the conclusions reached were the same: supernovae at redshift near 0.5 were about 25 percent fainter than they would be in an $\Omega_m = 1$ universe. The distant supernovae were, with a few exceptions where the teams helped each other out with observations, not the same. The data reductions were done by different methods. The ways that light-curve shapes were employed to correct for the variation in SN Ia brightness were different. We handled dust absorption in different ways. But despite these differences in detail, the conclusions were, as Saul neatly put it, "in violent agreement."

Although, as Gerson Goldhaber had correctly noted, we had fewer distant supernovae, 16 to their 42, on average, each of our points had about half the error. I think this was the good effect of having Nick Suntzeff as a leader of the high-z team plus the power of the statistical methods we had developed to analyze supernova light curves. The ability of a data point to tell you something decreases as the square of the scatter, so our 16 points with small

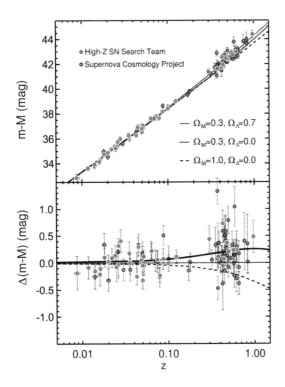

Figure 10.7. **The Hubble diagram for high redshift supernovae.** The small departure from the dotted line in the upper panel is the evidence that we live in an accelerating universe. In the lower panel, the 45° slope, which is just the inverse square law, has been removed. The points certainly lie above the downward curving line of long dashes, which is the prediction for $\Omega_m = 1$ with no cosmological constant. Most of the points also lie above the dashed horizontal line which is the prediction for $\Omega_m = 0.3$, with no cosmological constant. The only way to get up to the solid line (which is formally the best fit to the data) is to include the effects of acceleration. Points from both the high-z team and the supernova cosmology project are shown here. The high-z team points are fewer, but have equal weight because of smaller uncertainties.

scatter were just as helpful in telling something about cosmology as their 42.

And the something was, you needed Λ to match the data. Since there is an invisible contest between Ω_m, which slows cosmic expansion, and Ω_Λ, which speeds expansion up, the supernova results provide information about the difference between the attractive effects of matter and the accelerating effects of dark energy. The supernova results showed that acceleration is winning now, stretching out the distance light has to travel from a supernova at redshift 0.5 to our telescopes. The supernova results measure $\Omega_m - \Omega_\Lambda$, and they showed that this quantity must be smaller than zero. You cannot do that without Λ, or something very much like it. It's a little like stepping on a scale and finding your weight is *below* zero—something beyond the usual gravitational attraction must be going on! So far, the supernova data are the *only* evidence that the universe is accelerating, and the only measurement that shows the effects of Λ directly. As Sir Frank Dyson said of the gravitational bending of light, "I was myself a skeptic and expected a different result." Me, too.

The cosmological constant might have been Einstein's biggest blunder and part of Eddington's journey into the theoretical wilderness, but the evidence from supernovae shows that we need it, or something very much like it, to understand the world we live in. This is no longer a matter of esthetics or introspection or stubble from Occam's razor. We need to learn to live with Λ.

Of course, Brian Schmidt's horror made us take extra steps to be certain that the small extra dimming of distant supernovae was not due to some other effect. If somebody was going to find a flaw in this work, we thought it would be best if we did it ourselves. So we tried hard to see if we could show our own result was wrong, or misguided, or if we had missed some important source of error that was not described by the statistics of the data points.

We knew it wasn't Malmquist bias. Malmquist bias selects the brightest objects near the limit of a survey. But we weren't seeing supernovae that were extra bright, we were seeing objects that were extra dim. But there is more than one way to go wrong.

We know that when we look to redshift 0.5, we're looking back about one-third of the way to the Big Bang, about 5 billion years. So the stars will all be 5 billion years younger. Does age make a difference to supernova properties?

We know that the universe has grown richer in heavy elements, partly through the action of all the supernovae that have blown up in the past 5 billion years. Does chemistry make a difference to supernova properties near and far?

And we know for sure that many astronomical investigations have come to a bad end by misunderstanding dust. Couldn't boring old dust, not acceleration, make the distant supernovae appear dim?

These are serious questions to which the answers are still incomplete. Our job now is to examine these possibilities to see if they have misled us into the temptation of ascribing to cosmology an effect that truly belongs to evolving stellar populations or changing chemical composition or dirt.

As for the ages of stars, we know that galaxies today have stellar citizens with distinct demographics. Elliptical galaxies have very little current star formation, so all the stars are old, like the population of an Arizona retirement community. In contrast, spiral and irregular galaxies often have very active star formation—this is more like Ann Arbor, a town full of boisterous young people as well as a quiet older population. Those galaxies have young stars, including massive stars that blow up as SN II in much less than 5 billion years. They also have a quiet population of old stars that putter around while the young stars live fast, die young, and leave a beautiful neutron star corpse. So different types of nearby galaxies provide places to study the effects of a young or old population of stars.

Interestingly, type Ia supernovae have been found in all types of galaxies. It is worth looking to see if the SN Ia in spirals, where there is recent star formation, differ from the SN Ia in ellipticals where there is not. That would provide a clue to whether looking back in time makes a difference in the brightness of the supernovae. From the Calán/Tololo data plus the CfA data, we have now built up a set of over 50 well-observed supernovae in nearby galaxies that lets us examine this question. Every month we observe more. We are going to find out.

At first glance, the news is bad. On average, the SN Ia found in elliptical galaxies are dimmer than the SN Ia found in nearby spirals. However, when you use the light-curve shape to correct the luminosity, as we do for both the nearby and the distant sample, supernovae in ellipticals are indistinguishable from the supernovae in spirals. This suggests that there may be a real difference in the stars that become supernovae in spiral galaxies, and presumably in the distant younger galaxies we observe to measure Λ, but the correction methods we have developed are adequate to deal with this difference. By measuring the shape of the light curve, we iron out the age differences in the supernovae from 5 billion years in the past.

Does chemistry affect the brightness of supernovae, somehow making the supernovae in distant galaxies dimmer? There are several ways to approach this problem. Theory is one path. Peter Höflich, Craig Wheeler, and Friedel Thielemann wrote a paper in 1998 to look into the theoretical possibilities.[13] One prediction of supernova theory is that increasing the chemical abundances, as happens in galaxies over time, doesn't affect the spectrum or the overall light emission very much, except in the ultraviolet, where increased abundances are predicted to make SN Ia dimmer. This is the opposite of the effect we see, where the distant (and presumably slightly anemic) galaxies are *dimmer* than the nearby objects, which are the ones formed from enriched gas.

The chemical evolution from 5 billion years ago to today is not very extreme. In our galaxy, the chemical abundances 5 billion years ago at the site where the sun formed were precisely the solar abundances we see in the solar system today. Most chemical change in our galaxy and in other galaxies, took place in early violent episodes of star formation. Gas in our galaxy today, 5 billion years after the sun formed, is not much richer in heavy elements than the gas that formed the solar system.

Even so, it would be prudent to look for these effects. With the predictions of theory as a roadmap to action, we are now building up a sample of ultraviolet observations of nearby supernovae, since that's the part of the spectrum where chemistry matters most. We will see if galaxies with different chemical abundances produce SN Ia with different ultraviolet light curves and colors, as predicted. This work isn't finished, but so far, there do not seem to be large differences.

We have also compared the spectra of our high-z supernovae with the spectra of SN Ia observed nearby. The mighty Keck is amazingly good at obtaining spectra of the distant objects, using its immense collecting area to gather in the photons from distant SN Ia, and then sorting them out by wavelength. Alison Coil and Alex Filippenko led our group's effort to study the spectra of high-z supernovae and to compare them with supernovae in the local neighborhood. The spectra of distant SN Ia are, within our ability to measure, just the same as the spectra of the nearby SN Ia going back to SN 1972E and SN 1937C.[14]

The spectrum formed in the expanding atmosphere of an exploded white dwarf depends in a very complex way on the chemistry, velocities, and temperatures throughout the wrecked star. It is hard to imagine that exploding white dwarfs near and far are significantly different in light output, but have somehow conspired to make the spectra the same. Just because we can't imagine something doesn't mean it can't be true, but spectrum measurements test whether distant supernovae are distinctly different from nearby ones. If so, the cosmological interpretation of the supernova results would be suspect. This is a test SN Ia could have failed, but as far as we can see, they did not.

Dust is trickier. We know how to detect the presence of dust like the dust in our galaxy from the reddening it produces. Adam Riess worked that out in his Ph.D. thesis, and the Tololo crew did something equivalent. We made all our measurements of high-redshift supernovae in two colors specifically to overcome that weakness in the earliest SCP data. But clever theorists can invent dust that is unlike the dust in the Milky Way, and perhaps pixie dust like that really exists. A Harvard astronomy graduate student with a slight contrarian bent, Anthony Aguirre, worked this out. As a beginning student, Anthony had examined the possibility that the microwave background wasn't really from a hot Big Bang, but

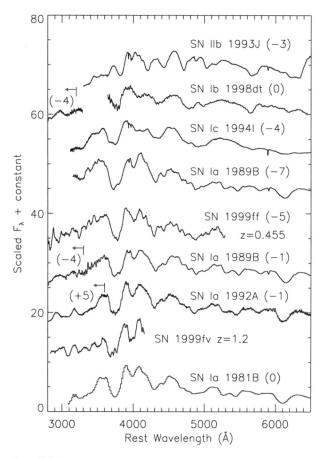

Figure 10.8. **Spectra of supernovae**. The supernovae observed at high redshift, SN 1999ff and SN 1999fv are, as near as we can tell, identical with those seen nearby at similar ages. The spectra have been shifted back to the wavelengths you would observe if you were in the same galaxy as each of the supernovae. Courtesy of Alison Coil, Alex Filippenko, and the High-z supernova team.

might be thermal emission from solid particles. Again challenging orthodoxy, Anthony asked whether there could be pixie dust in the universe that dims the light of distant supernovae, but does not leave the fingerprint of reddening. To explain the high redshift supernova results this way, you need dust that dims distant supernovae by about 25 percent and that eludes our color measurements. Is this possible?

Anthony knew that the effect that leads to reddening has its origin in the sizes of interstellar dust particles. When the particle size is comparable to the wavelength of the light waves, you get reddening. Interstellar dust is not like the big balls of dog hair and sloughed-off human skin that accumulate behind your couch. Interstellar dust is a very fine submicroscopic haze made of carbon and silicon that would be invisible to an ordinary microscope. It is a kind of soot and sand that even the finest white glove test would not reveal. Although that's the kind of dust we know about, Anthony suggested there could be another kind of dust that we don't know about, which is revealed only in the supernova data. These would be big dust grains, so large that they affect all colors almost equally. Big dust grains could lead to absorption without much reddening.

Scientific imagination has to obey reason and cannot violate observed limits. Anthony had to think how to make the dust smoothly distributed throughout the huge regions of space between the galaxies. Otherwise, the light from a supernova would sometimes encounter pixie dust and sometimes not. This would lead to increased scatter in the high-z supernovae, beyond what we observed. So he invented a story, which was not too crazy, for a way to form the dust, make sure it had big grains but no small ones, and expel it from galaxies. He had to be careful not to use more of the element carbon than stars could produce. Anthony found that there was a possibility that this specially constructed dust might exist, and showed that it could conceivably account for the results we were observing in the high-redshift supernovae. While there was no other evidence for this pixie dust, it was not impossible.

If this were a debate, you could ask, in a rhetorical flourish, pounding the podium: "I ask you my friends and fellow countrymen, which is more likely, a form of intergalactic dust that has hitherto eluded detection or creating out of whole cloth a mysterious new component of the universe—so-called dark energy that purportedly dominates cosmic expansion? Must we repeat Einstein's notorious stumble? Can we not learn from the past? Must we stride confidently into the abyss of error?"

Luckily, science does not consist of public exhortation. I encouraged Anthony in his work. Of course, if he was right, our measurements were about my least favorite subject, dust, rather than detecting a dramatic new component of the universe that had been hidden from observers for 80 years. But since we're trying to get at the truth, we should test every link in the chain of inference. Though debate can be amusing, to test Anthony's idea we needed a more sensitive measurement for the existence of his pixie dust.

First, Anthony improved his earlier prediction of exactly what this dust would do. On closer inspection, it was not perfectly gray, dimming all wavelengths equally, but slightly pink, absorbing blue light a little more than red light. To look for this, we have been observing distant supernovae over a larger span of wavelengths, where these subtle effects should show up. In practice, this means getting observations from the blue end of the emitted spectrum out to the infrared wavelengths that lie beyond the range of human vision. So far, we have one well-observed case, SN 1999Q, and there we see no sign of pink dust. In the year 2000, we observed a number of supernovae as carefully as we could over a wide wavelength range, to see if we can put a stake through the heart of gray dust. We put the data in the capable hands of my graduate student, Saurabh Jha. We shall see what he comes up with.

So we have some evidence that the age of the stars, their chemistry, and pixie dust are not the cause of the effect we see: the distant supernovae haven't been shown to be faint for one of these reasons. But we have a stronger way to distinguish a genuine cosmological effect from a misleading systematic effect in the supernovae. If we imagine, for a moment, that the apparent faintness of distant supernovae with redshift $z = 0.5$ is due to cosmology with a dominant dark energy causing acceleration, we can predict what will happen as we look even further into the past.

The brightness we measure for a supernova depends on the outcome of a battle between the accelerating effect of Ω_Λ and the decelerating effect of Ω_m. What we see is that dark energy has been winning this tug-of-war in the last 5 billion years or so while the light from supernovae at $z = 0.5$ has been on its way to us. But what about even more distant supernovae?

If we look into the past of the expanding universe, each chunk of it would have been smaller in the past. Imagine a region, say 500 million light-years on a side, that has stretched out from a smaller volume in the past. The amount of matter in such a big chunk of the universe has not changed appreciably over that time—that's too big a region to have been affected much by the individual motions of galaxies or the growth of structure. So if you look back to redshift of one, each of the sides of a cube that expands along with the universe was smaller by a factor of two back then. If you have the same amount of mass in a smaller volume, that means the density was up by a factor of 2^3, a factor of eight, so we would be looking back to a time when the universe was eight times denser.

When you look into the past, you see a denser universe. As you look further back, Ω_m gets more important compared to Ω_Λ. Einstein guarantees that if the total Ω is 1, it stays 1, but the balance between Ω_m, tugging to slow things down, and Ω_Λ, urging the universe to accelerate, ought to shift as you look deeper into the past. The distant past would be dominated by dark matter, not by dark energy.

If you think of the universe starting out from the Big Bang, the first several billion years would be sluggish years of slowing down due to gravity from dark matter, but then, as matter thins out and its density declines, the balance would shift. Dark energy, negligible at first, is destined to dominate cosmic evolution. Slow and steady wins the race. There will come a time, which depends on the precise values of Ω_m and Ω_Λ, where deceleration due to dark matter loses its grip and acceleration begins.

What makes this story interesting is that we can already see back to the time when the brakes came off and the gas pedal was mashed to the floor. For reasonable values of Ω_m and Ω_Λ that are consistent with the supernova data we had in 1998, such as $\Omega_m =$

0.3 and $\Omega_\Lambda = 0.7$, the coasting point that marks the transition from a decelerating universe to an accelerating one is at a redshift of about 0.7. This is distant, but not more distant than observations we've already made. As we look further into the past, we should see less acceleration from dark energy and an increased influence of dark matter slowing things down. And the redshift where this happens is not out of reach.

All the other effects, such as the age of the stellar population, the chemical composition of stars, and absorption by gray dust would *increase* with redshift. After all, if the ages of the stars matter, looking to bigger redshift means you are looking at younger stars. If composition matters, as you look farther back, you should see even lower abundances. And if pink pixie dust pervades the universe, you will traverse more of it looking back to higher redshift.

This makes a definite prediction—if the dimming we see is due to cosmology, then if we look far enough back, back past the coasting point, the extra dimming will grow smaller, and then shift sign to produce extra brightening. On the other hand, if other effects of age, composition, or dust are the cause of distant supernovae appearing dim, then you'd expect more distant supernovae to be extra dim. So we have a way to tell if we've been fooled or whether the effect we see really is due to the history of cosmic expansion. Look at higher redshift.

To observe the signature of dark energy, we need to find and measure supernovae out beyond a redshift of one. If they turn out to be extra dim, we're wrong and something that is not cosmology is making the supernovae dim. If they are brighter than you'd otherwise expect, this means we're on the right track. To find out, we needed to shift our attention from $z = 0.5$ to $z = 1$ and beyond.

By 1998, we were out on a limb, and we had the means to saw it off ourselves. Better us than somebody else. Worse than the risk of failure was the creeping sense of investment. While everybody on the high-z team was uneasy at first about a claim that Λ was real, as our own evidence built up, and the data from the other team gave the same result, we began to get comfortable with Λ and then slowly we began to like it. We began to care what the universe was like. This is not altogether healthy. You like to think you're just a dispassionate observer—an eagle-eyed umpire: "I call 'em the way I see 'em." But we were drifting into becoming believers, or at least fans of the cosmological constant. It was a time for less propaganda and more data.

11

the smoking gun?

By the beginning of 1999, the case for an accelerating universe had acquired some traction. The high-z team results were published in the *The Astronomical Journal* and the Lawrence Berkeley Lab results were widely circulated as preprints and conference reports, both pointing to the same conclusion. Consistent results from two independent teams made the evidence more credible, though there was still an outside chance we were both making identical mistakes. But if we were making a mistake, it was a subtle one. As far as either team could tell, distant SN Ia were just like the nearby SN Ia. Although age, composition, and dust were possible complicating factors, their effects seemed to be small, and we were working hard to detect and limit them. The decisive test, though, would be to look for supernovae at even higher redshift.

For any combination of the dark energy, Ω_Λ, and the dark matter density, Ω_m, we could predict what to expect as we looked to redshift one and beyond. After the Big Bang, there should be cosmic deceleration caused by Ω_m followed by cosmic acceleration due to Ω_Λ. Somewhere out beyond redshift one, supernova observations should reach well into the deceleration zone, and we should start to see the supernovae appearing a little brighter than otherwise. If the supernovae showed extra dimming instead, that would point to a systematic problem and we would be forced to retreat

18 December 1998

Science

Vol. 282   No. 5397
Pages 2141–2336 $7

THE
ACCELERATING
UNIVERSE

Breakthrough of the Year

AMERICAN ASSOCIATION FOR THE ADVANCEMENT OF SCIENCE

Figure 11.1. The work on the accelerating universe was Science Magazine's "Science Breakthrough of the Year" for 1998. Reprinted with permission from *Science*, vol. 282, December 18, 1998. Illustration: John Kascht. Albert Einstein™ represented by the Roger Richman Agency, Inc., Beverly Hills, CA. Copyright 1998 American Association for the Advancement of Science.

from our claim that Λ is real. Aside from the embarrassment, there would be regret. We were beginning to like Λ.

And we weren't the only ones. *Science*, the leading science journal in the United States, selected the accelerating universe as its top "Science Breakthrough of the Year" at the end of 1998. Brian Schmidt won Australia's first "Malcolm McIntosh Prize for achievement in the physical sciences." Brian was in demand, at least in Australia, where the Australian Broadcasting Company called him in for an interview on the "The Age of the Universe" show, hosted by John Doyle, known to a wider audience as the TV announcer in the movie *Babe*.

For another voice, they asked me to come to a TV studio at 11 P.M., which was convenient for the Australians. Jayne and I had tickets to the Red Sox game at Fenway. Stupidly, I drove so I could get over to WGBH after the game, and even more stupidly I parked in one of those lots marked "E-Z Out" where they block your exit with as many cars as possible after you have left the scene. The Cleveland Indians and the Red Sox were engaged in a titanic struggle as the clock ticked past 10:30. Relief pitchers paraded in for both teams. Contrary to the oath I had sworn to my grandfather in the Fenway bleachers in 1959, I needed to leave the game before the last out to be on Australian TV on a program hosted by a comedian. My car was wedged in the parking lot. I backed and filled while Jayne gestured with an imaginary steering wheel, slowly ooching around to the angle where I could escape. Then I heard a sickening sound of metal bending, as I warped the curve of my Saab's door on the bumper of a Chevrolet. There was no going forward and there was no retreat. At that instant, I realized it is better to do the work than to talk about it on TV.

Plenty of people were working on the accelerating universe. There was a flood of theoretical papers concerning the dark energy, which was quickly seen to be a frontier of physics.[1] The idea that the universe was composed principally of vacuum energy with negative pressure, required by these astronomical observations but nowhere seen in terrestrial laboratories, meant that an important problem in basic physics was not yet solved. Presumably, this has something to do with the fact that there is not yet a complete "theory of everything" that treats gravitation on the same quantum footing as electromagnetism, the weak nuclear force, and the strong nuclear force. Effects of virtual particles and their antiparticles that spring out of the vacuum and annihilate one another are staples of theoretical physics for electromagnetic effects. Amazingly enough, the Casimir effect and the Lamb shift are both laboratory experiments that show these wild ideas have real consequences that agree with the facts. The vacuum is a lively place.

For the gravitational equivalent of vacuum effects, there are as yet no laboratory experiments and no well-established theory, just the evidence from supernovae for an accelerating universe. But there are adventurous ideas. My union brothers at the Institute for Theoretical Physics had been among the pioneers of string theory, which people tell me works best in a bulky 11-dimensional space. There is some hope that an explanation for the small value of the cosmological constant in our membrane of three space dimensions and one time dimension might drop out of this cogitation. So I have been skimming the abstracts of papers with titles like "A Scalar-Tensor Brane World Cosmology." Despite holding a union card in the International Brotherhood of Theorists, my true orientation is much like that of Hale: "I confess the subtleties of the theory are altogether too much for my comprehension." Astronomers learn general relativity now, so Hale's modesty seems quaint. Someday we may have to understand 11-dimensional M-theory or its descendents to understand cosmology, though for now the ferment is among "the very few competent to discuss the matter with authority."

The idea of looking for the fingerprints of dark energy in supernovae beyond a redshift of one was not ours alone. The Supernova Cosmology Project was still months ahead of us in many ways, and they had already found one, SN 1998eq. During the interim before supernovae are reported to the International Astronomical Union to receive their designations, each team uses its own nicknames for the candidates, a little like the names for hurricanes. Saul Perlmutter is a cultured fellow, a Harvard graduate and a violinist. Their team decided to call their really high redshift candidates according to an alphabetical list of composers, starting with Albinoni.[2] So far, the data on Albinoni, which is said to be at a redshift of 1.2, have not been published, though in talks they show it lying below the line you'd expect if misleading effects like age or dust are most important, and in the general direction for a genuine cosmological effect.

During 1999, John Tonry of the University of Hawaii led the charge toward higher redshift for our high-z team. John is a creative astronomer who perfected a new way to find the distances to galaxies based on how grainy they look. He is an independent guy and a software wizard of the first magnitude. John has deftly constructed his own version of exquisite sky-subtraction software and reexamined the whole Rube Goldberg scaffolding of the high-z team's way of doing things. This incredible effort not only protects us against a missing π or minus sign in the computer code, but has improved our ability to find supernovae at redshifts of one and beyond. As you look for supernovae at higher redshift, you not only must find fainter objects, you must look further to the red, because that's where the light from very distant supernovae is shifted. Then you have to deal with more light from the sky, which glows in the near infrared, and with CCD detectors that generally do not work as well at those wavelengths as they do for observations at visible wavelengths

So everything is working against you: the objects are fainter and redder. Because they are redder, you have to use detectors where they are less effective, and you have to contend with a brightly shining sky even when the moon is not a factor. But the reward for doing these difficult observations would be to find supernovae that reveal the autograph of Λ. It seemed worth the effort.

To tune the search for higher redshift, we changed tactics. Since the target supernovae were going to be fainter than the ones we had previously sought, we made our exposures longer. Since they were going to be at higher redshift, we shifted the filter of our exposures farther to the red. And since the sky was more of a problem, we really needed the improved software that Brian Schmidt and John Tonry developed to subtract one frame from another. In keeping with our policy of incorporating every good idea we can find, we improved our image subtraction by incorporating a scheme developed by Christoph Alard into our data pipeline.

On 2 and 3 November 1999 our team discovered 20 supernovae using the giant $12{,}000 \times 12{,}000$-pixel CCD array camera on the Canada–France–Hawaii Telescope by subtracting images taken the month before. Confirming images were taken at Cerro Tololo, using the 4-meter Blanco Telescope. Spectra obtained in the next 10 days at the Keck by John Tonry and by Alex Filippenko showed that 12 of our candidates were SN Ia. Two of these proved to be especially interesting targets with redshifts greater than one. To provide a little contrast with the other team, we took a lowbrow path and named our candidates after cartoon characters instead of cultural figures.

So we had Rocky and Bullwinkle, Boris and Natasha, and Fearless Leader as candidates. The ones that were most interesting were Velma and Dudley Do-Right. Although measuring each spectrum was tough going, these appeared to be at redshift 1.05 and 1.2.

In December we continued to follow the decline of these supernovae to measure the shape of the light curve. At high redshift, time is stretched out by expansion just the way the wavelengths of light are stretched. This means that in a month of our time, the supernova only ages by two weeks. So our December observations were well placed to see what the supernova was doing in the first two weeks after maximum light. Since I was in Hawaii anyway for my honeymoon, having just gotten married during the full moon in December, I went up to the summit of Mauna Kea to observe with my graduate student Saurabh Jha. My wife, Jayne Loader, compensated for the psychic injury by having her toenails painted at the Mauna Kea Beach Hotel.

Saurabh had already mastered all the details of observing at the University of Hawaii's 2.2-meter telescope on Mauna Kea. At sunset, the mountain is a fabulous place: an extinct (we hope) volcanic landscape without vegetation. The clouds of the trade winds were below us, and above us only 60 percent of the air that Jayne was breathing at sea level. By midnight, the absence of air was beginning to bother me. Legend has it that you can't think straight at the 13,796-foot summit. But if you can't think straight, why should we believe what you say about how you are thinking? In any case, I was getting a headache, my gums were sore, and I felt a little short of breath. But Saurabh seemed alert and we were getting excellent data. It was fun. If you're running the show, you get to choose the music, so Saurabh was playing an austere minimalist composition by Steve Reich on his CD player. As the rhythmic figures subtly wove into patterns, I tried to ignore my aching gums and to think about Λ. We were working at a high enough redshift to see beyond the era of acceleration, back to the time when matter ruled the universe. I could see the galaxies forming in dark matter lumps as they rushed outward from the Big Bang, the whole expansion slowing due to the tug of dark matter, and then the steady push of the cosmological constant shifting the balance and driving space outward faster and faster, fading to red and leaving us alone in the darkness, gasping for breath. Or maybe I was just oxygen-starved. When I awoke, the integration was done, and it was time to shift to the next target.

The University of Hawaii tests your mettle by making you observe at the summit, but the Keck Observatory compensates for the effects of altitude by spending money to let the observers work at sea level. Since you never need to touch the telescope anyway, controlling the instrument through a computer console, you might as well get some oxygen. Technicians at the summit babysit the telescope, but the observers are down in Waimea, using a fast computer link to control the data-taking instruments. If they were on the summit, they would also be in a control room, using identical computers, so there's not much to lose and a lot to gain in mental alertness and physical comfort. This gives you more chance to choose your next move intelligently, taking into account the weather, how sharp the images are, and your list of targets.

There's a slow-scan TV link that lets you observe the telescope operator sitting patiently in a chair, and for the operator to see you, too, frantically trying to calibrate the latest observation fast enough to decide what to do next. That communication is good enough. You don't become close friends with Wayne the operator, encountering him only in this distant way, but that's a reasonable price to pay for a brain that works.

Sometimes strange things happen—one night it was pouring down Hawaiian rain in Waimea while we were taking spectra of supernovae. Up on the summit, in Wayne's world, they were above the moisture of the trade winds and observing conditions were excellent. As we were debating which object to do next, Barbara Schaefer, the head of all the Keck telescope operators, looked into the data room on her way home from a late night at the office. We described what we were doing—going after redshift one supernovae. We were trying to decide whether to do another of these nearly impossible objects or to do something easier where we would be sure to get a useful result. Barbara composed her face in serenity, placed her palms together, and, in the cheddar-sharp nasalities of upland Wisconsin intoned the koan of Keck: "When conditions are good, do the hard thing." Then she went home to her cats.

The logical extension of this mode of observing will be to use fast network connections so you can observe in Hawaii without leaving Berkeley or Pasadena or Cambridge. Instead of amusing the scantily clad tourists and locals by arriving in Kona with a down parka and heavy boots for the summit, or sleeping through a perfect day in Waimea, you will be at your office with the phone ringing, and students waiting outside the door, where the only respite will be the chance to sleep through faculty meetings. This will be followed by a long night of observing. We will call this progress.

There are many large new telescopes coming into operation, including the twin Gemini 8-meter telescopes on Mauna Kea and in Chile. To put the Gemini Observatory on the scientific map, they organized a conference on "Astrophysical Timescales" and invited me to speak. It seemed like a great place to discuss how Λ affects the estimate of time elapsed since the Big Bang, so I said yes. But when it came time to plan the travel, I realized there were other timescales I could not alter. There was no way to get to Hilo, Hawaii and back without missing a lecture in my undergraduate class at Harvard. They don't ask us to teach all that much, so I try to be there every time. Instead, John Tonry hopped over from Honolulu to report on the results from our high-z observations at the Gemini conference. This was better, and not just because of logistics: John had been doing most of the work and it seemed right that he should give the talk.

Although the final analysis was not quite done and John's conference report has not been refereed, so it doesn't have the weight of a real journal article, it does show which way the finger of fate is pointing. If the distant supernovae were yet fainter, it would be bad news for Λ. If the distant supernovae came out a little brighter than that, it would be the signature of cosmic deceleration in the early universe, and a clear sign that the effects we were observing were cosmological, not the result of "a changing population of supernovae" feared in an ancient News and Views. John showed that the data from Dudley Do-Right, at $z = 1.2$, and his high-redshift friends, came out a little bit brighter than you'd otherwise expect. In the context of a cosmology with Λ, our new data favor a universe that is accelerating now, but was decelerating in the distant past, 7 billion years ago. This stop-and-go universe is good news for dark energy.

Since the Gemini meeting in Hilo was a conference on cosmic timescales, John also spelled out what Λ means for the connection between cosmic expansion and cosmic age. If you have a Hubble constant of 72 kilometers per second per megaparsec, then $1/H_0$ is 14 billion years. If there were no acceleration and no deceleration, that would be the real elapsed time since the Big Bang. In an $\Omega_m = 1$ universe completely dominated by dark matter, the deceleration of the universe means that the present rate of expansion is lower than the average, so the universe is younger than it currently appears, with an elapsed time of about 9 billion years. This conflict with stellar ages of around 12 billion years was one of the rhetorical gambits advanced for Λ before the observational evidence from supernovae.

In an accelerating universe, the real age could be larger than the apparent age, but in a stop-and-go universe, as suggested by the high-z data that John Tonry presented, it could go either way. If the slowing down were more important, the universe would be younger than 14 billion years, and if the speeding up were more important, the universe would be older.

By coincidence, if $\Omega_m = 0.3$ and $\Omega_\Lambda = 0.7$, which is a good representation of the supernova data, including our new points at redshift one and beyond, then the slowing down and the speeding up just about balance, and the elapsed time from the Big Bang to now, is just 14.1 ± 1.6 billion years for a Hubble constant of 72. So, after all this lucubration, including cosmic deceleration followed by cosmic acceleration, it looks like the answer you get using third-grade arithmetic is the right answer for the age of the universe. And that answer is in good accord with the ages of objects in the universe. So far, so good.[3]

But if going to redshift 1.2 is good, wouldn't it be better to go even farther into the past? The effects of deceleration would be larger, the contrary effects of pixie dust would be larger, and the difference between them would be even more impressive evidence that we were seeing cosmological effects based on the history of cosmic expansion, not illusions caused by stellar ages, chemistry, or absorption. But John Tonry had already led us pretty close to the limit of what can be done from the ground: we were using the world's largest telescopes at the world's best sites. The next step needed to be taken above the Earth's atmosphere.

Although the Hubble Space Telescope is not good for a wide-angle search, it has no peer for peering deeply into a little patch of sky. In 1995, Bob Williams, then Director of the Space Telescope Science Institute in Baltimore, promoted a project to stare at an otherwise blank and uninteresting piece of sky with the Hubble Space Telescope. He consulted widely to be confident that this "Hubble Deep Field" had broad community support. There is just one space telescope and, though the director is responsible for setting the scientific program, and can nominally do what he thinks best, in practice the "director's discretion" time is limited, and most of the space telescope's observing program for every year is decided in a bruising peer review by a Time Allocation Committee.

The previous Director, Riccardo Giacconi, had used his discretion to deal with scientific opportunities that cropped up between cycles, or to rectify injustices caused by the Byzantine rules of the time allocation process. But Bob Williams wanted to concentrate his Director's time on a single spot to drill the deepest well into the past that technology would allow. Personally, I thought it was a dumb idea.

In an expanding universe, the contrast of galaxies with the sky should fade out in proportion to $(1 + z)^4$, so once you got beyond z of one, you were losing a factor of 2^4, which is 16, and you were running the risk of seeing nothing much. Why invest so much valuable telescope time on this potentially futile effort when you were turning down good proposals (including some of mine) every year?

Fortunately, Bob received many opinions, not just mine, and he went ahead. The Hubble Deep Field observations produced a gusher of information on the past—especially the past history of star formation in galaxies. The distant galaxies were not only visible against the night sky (because they were lumpy—who knew?) but they were just within reach of the Keck telescopes for spectra, so the Hubble Deep Field is not just a knockout screensaver, but a powerful window into the history of star formation in the universe at redshift 1, 2, 3, and beyond.

In 1996, Ron Gilliland of the Space Telescope Science Institute and Mark Phillips, from Cerro Tololo, applied for time to revisit the Hubble Deep Field. A second look would allow them to detect things that had changed. A second look would allow them to find supernovae at high redshift. When Mark described what they had planned, I was lukewarm. A back-of-the-envelope calculation showed that their chances of finding anything were not very good, and even if they did find something, without an extensive follow-up program, they weren't going to learn very much. Without a light curve, they wouldn't know if the supernova was going up or down. Without a spectrum, they wouldn't know the supernova type or the galaxy redshift. And they were looking in the wrong place for light from a very distant supernova: a supernova beyond redshift 1.5 would have its emission shifted out into the infrared, beyond 1 micron, where the CCD detector on HST was completely indifferent to light. On the other hand, there wasn't much harm in trying, and they might get lucky. Fortunately, I wasn't on the Time Allocation Committee that year and they got the time.

In data from the repeat observation, which comprised 18 orbits exposed between 23 and 26 December 1997, scrupulous subtraction by Ron Gilliland showed there was a Christmas present for Ron and one for Mark. There were two definite dots: SN 1997ff and SN 1997fg. Ron and Mark, together with Peter Nugent, wrote up their discovery for *The Astrophysical Journal*.

On one hand, I was right. They didn't have enough information to do much with this detection, and it didn't add much to the cosmological story that was unfolding. On the other hand, they were right—they showed that HST can be used to find very faint new objects. And more than anyone knew at the time, we had all been very, very lucky because SN 1997ff was about to be observed again and again in just the right way to add to the story of the accelerating universe.

Earlier in 1997, when astronauts rode the Space Shuttle up to HST, they carried two new instruments—a much improved spectrograph called STIS, and an infrared camera called NICMOS. While having the space telescope in a low Earth orbit creates huge headaches for planning observations, it does make it possible to bring up new instruments for a telescope that had been designed in the 1970s. NICMOS was a little infrared array, something like a CCD, but with a light-detecting ability that extended out to 2.5 microns in the infrared, roughly five times the wavelength of visible light. Infrared emission comes from cool places and infrared light is not obscured by dust as much as visible light, so NICMOS was a powerful tool for probing the cool, dusty places where stars are being born.

Compared to the CCDs on the Space Telescope, the NICMOS array is very modest—it has only 65,000 pixels, compared to 2.5 million for the visible-light camera. It covers a tiny patch of sky smaller than 1 arcminute on a side, while the CCD array covers an area 8 times larger with finer pixels. But it had one powerful new property—with sensitivity in the infrared, NICMOS looks at wavelengths where very distant galaxies and supernovae are brightest. Observing from space gives sharp images, but even more importantly for infrared observations, there is much less emission from the sky. As a result, NICMOS can knock the socks off 10-meter telescopes on the ground at the job of measuring infrared light from distant stars and galaxies.

Ordinary stars in high redshift galaxies emit visible light that gets redshifted by cosmic expansion into the infrared. So it seemed to the NICMOS team like a good idea to follow up on the success of the Hubble Deep Field, which had been done in visible light, by using NICMOS to pound away for 100 orbits on a tiny patch to see what would show up in the infrared. After a test exposure on 26 December 1997, the NICMOS team started their observations in earnest on 19 January 1998. Without the intervention of human intelligence, by pure good luck, SN 1997ff was in the corner of their small field of view, like a hummingbird in a family snapshot on the fourth of July.

In science, as in life, it is good to be lucky! To build up a deep field, the NICMOS team returned again and again to the same place, slowly accumulating more and more data to beat down the noise and to allow the faint galaxies to be seen. Over a period of 32 days, HST accumulated many exposures of the same place. And almost every one had an infrared image of SN 1997ff, building up the material for a beautiful light curve for this object. But nobody knew that. The data went into the STScI Archive, where they aged like a fine claret from Bordeaux, just as Bev Oke had placed his supernova spectra in his Caltech desk to ripen for the opportune moment.

Last year, I was visiting the Space Telescope Science Institute as a member of one of the myriad committees that provide sage advice to NASA on how to proceed. We were discussing the Next Generation Space Telescope. NASA had learned how to put a man on the Moon: by using checklists. Their confidence in the efficacy of paperwork has now been transmogrified into a worship of landscape-format Powerpoint presentations. To escape from the blizzard of charts during a coffee break, I walked down the hall to see my onetime student Adam Riess.[4]

Things were going well for Adam. After getting his Ph.D. at Harvard, he had gone on to become a Miller Fellow at Berkeley. His thesis on SN Ia had won the Trumpler Award for the best recent Ph.D. He had married Nancy Schondorf. He was the first author of our paper on the accelerating universe. He now had a real tenure-track job at the Space Telescope Science Institute. He won Harvard's Bok Prize, awarded to one of our astronomy department graduates for outstanding work before the age of 35. Alex Filippenko and I risked indictment for perjury in writing Adam incandescent recommendations for the American Astronomical Society's Warner Prize. He won that, too. His picture was in Time Magazine, pleasing his mother no end. The Warren, N.J., *Echo-Sentinel* ran a front page story under the headline, "Local Boy Does Well in Astrophysics." He was buying a house. And now, he had a really great result to show me.

"Don't tell anyone," he told me, carefully closing his office door. "I'm still working on this. Wait 'til you see!"

Inwardly, I was chuckling. Adam was the gossip, not me.

While the meeting droned on down the hallway, Adam showed me the graphs and pictures that told the story. Given the quality of my advice to NASA over the years, my absence from the meeting may have been a net benefit to society. Adam had scoured the HST archive and dredged up the NICMOS data for SN 1997ff. The rule is that you have one year for proprietary use of your own data, but then everything becomes public. STScI had built an excellent archive and encouraged people to exploit it. The statute of limitations had run out for the NICMOS team. Anyone could do what Adam was doing. Maybe somebody else was.

"Don't tell anyone."

"Get on with it."

With the Hubble Deep Field itself as the "before" image and the discovery data from Gilliland and Phillips at visible wavelengths, the repeated NICMOS observations made a fantastic data set. Supernova 1997ff was a bigger, fatter dot on the infrared images than in the CCD data. Adam was working with a whole squad of competent people who knew the details of the Hubble Deep Field and of NICMOS, including Rodger Thompson, leader of the NICMOS team. Adam was putting it all together into an amazing picture for SN 1997ff. They had a great light curve. They had color measurements. They had good data about the properties of the galaxy in which the supernova had gone off. It looked like an elliptical—the kind that has only SN Ia and little dust. The observations also yielded a redshift for the galaxy estimated from its colors. Adam showed that you could, independently, get a redshift estimate for the supernova from its colors. The two methods agreed. The redshift was about 1.7. This was what we had all been dreaming about doing—and the data had already been gathered without any planning or filling out of forms.

The payoff was to see whether this supernova from deep in the past could tell us whether we had made a colossal mistake in drawing the inference of an accelerating universe. Adam got slower and slower in flipping through the figures. He was enjoying the suspense. Even more, he was enjoying the turned tables. How many times had I revised his manuscripts and been the one holding the authority to sign off on a result, to approve his thesis? No more. A student had become a colleague, and he was relishing it.

Was the supernova dim, showing we had been fooled, or bright, pointing the finger at Λ? Adam kept the final chart face down.

"If this thing turns out dim, Adam, you'll have to give back the Trumpler Award and the Bok Prize. Nancy will have to decide for herself what to do with you."

I was joking, but I was burning with impatience to see him flip over that ace in the hole.

"Look at this."

Supernova 1997ff was extra bright for its redshift, the way it should be if the universe was decelerating at first, then accelerating.

"Adam, this is really good."

"I know."

"You are really, really lucky—the NICMOS guys could easily have chosen another spot to observe."

"I know."

"Adam, this is really good and it is really important. The NASA press machinery will lap this up. But don't believe everything you read."

On 25 June 2001, Mike Lemonick at *Time* wrote up the story of SN 1997ff and, in a week without a terrestrial disaster, they ran it on the cover: "How the Universe Will End." Since they already had a file photo of Adam, they used it in a photo sequence his mother loved even more. It showed Einstein, then Hubble, then Zwicky, then Penzias and Wilson, then Adam Guy Riess, explorers of the cosmos. I told Adam they were arranged in decreasing order of importance but increasing lovability.

This result is too important to rest on just one object, but SN 1997ff points toward a universe that is genuinely a mixture of dark matter and dark energy. Further observations with HST will reveal more of those very distant supernovae and show more clearly whether we live in a universe that was slowing down before it began to accelerate. That will be the smoking gun for Λ. But SN

1997ff is a test that the accelerating universe could have failed. And it did not fail that test.

Supernovae are the only direct evidence for acceleration, but shortly after the first supernova results in 1998, we began to combine the supernova data with observations of the ripples in the cosmic microwave background that can determine the cosmic geometry. Martin White, as a postdoc at the Center for Particle Astrophysics at Berkeley, again when he was at Illinois, and later as a colleague on the Harvard faculty before he defected back to Berkeley, had repeatedly pointed out to me that when experiments measure the angular scale of the freckles on the microwave background, they measure the cosmic geometry. You could learn the total Ω. When combined with the supernova observations, these measurements pin down how much dark energy and how much dark matter the universe contains.

Here's the way that works. The era when the universe was opaque ended about 300,000 years after the Big Bang. So the biggest scale of temperature variations that result from variations in the matter ought to have a size of something like 300,000 light-years. This is similar to the biggest variations in water level you can make in a bathtub—the longest waves are the size of the tub. You can try this at home. If anybody objects to the mess, you just say you are studying the formation of acoustic waves in the early universe. We view those ripples at a distance of 14 billion light years. Now, the angle covered by an object of known size (300,000 light-years) at a known distance (14 billion light-years) depends on the geometry of the universe. And Einstein tells us that matter and energy, the total Ω, that is Ω_m plus Ω_Λ, determines the curvature of space. If the universe has the geometry of a sphere (Ω greater than one) the freckles of the CMB will cover a larger angle than if the universe has the geometry of a saddle (Ω less than one) or has the flat geometry predicted by inflation (Ω exactly equal to one).

The supernova data gave a value for $\Omega_m - \Omega_\Lambda$, and the microwave background gives a value for $\Omega_m + \Omega_\Lambda$. Even a person whose theoretical credentials were fabricated in Photoshop could see that this would let you measure both Ω_m and Ω_Λ. If you know that the sum of the ages of Becky and Bob is 79 and you know that Bob's age minus Becky's is 26, that's enough to tell how old each one is. If you have measurements from supernovae and from the microwave background, you can learn how much dark matter and how much dark energy makes up the universe.

In 1998, measurements of the fluctuations in the microwave background were just beginning to provide credible measures of the total Ω. Back in 1992, the COBE satellite had shown that there were fluctuations, but the COBE map of the sky was a blurry one, averaging over patches about $7°$ on a side. Small-scale freckles would be smoothed into patches of tan by those measurements. The recent balloons and ground-based systems were designed with the acuity to see variations on the scale of one degree or less.

The experimenters were working like demons to get their measurements before results from the next satellite, Microwave Anisotropy Probe (MAP), dominated the field. MAP was specifically designed to make a fine angle map of the microwave emission. But as with any satellite, between the design and the launch, technology marches on. The detectors in MAP were conservative designs when they were thought out, over 7 years before the measurements. Agile balloonists, dwellers in high deserts, and Antarctic adventurers use more recent technical developments to detect the microwave signals. Though their observing sites have to contend with more interference from the Earth's atmosphere, if they are very clever and somewhat lucky, the little mice can sometimes do very well compared to a lumbering elephant of a space project.

In 1998, the situation was confusing, but hopeful. There were many measurements that suggested roughness in the cosmic background at an angular scale of about one degree. But some measurements disagreed with others and there was no single set of measurements that showed by itself that this signal was there. For outsiders to the field, like our team, it was hard to know how to use this information. Stepping up to the challenge, some energetic workers set themselves up as knowledge brokers, combining the data from various experiments to extract something reliable from the conflicting evidence. The supernova data and the CMB data were complementary. They defined two lines, perpendicular to one another.

Where they crossed, X marks the spot, and that's where the treasure was buried. Ω_m and Ω_Λ. Our kind of treasure.

My postdoc, Peter Garnavich, was eager to plunge into this game. Peter was a late bloomer. He had been a graduate student at MIT, then quit and worked at the Space Telescope Science Institute before going to the University of Washington for a Ph.D. He had observed exactly one supernova as a graduate student, but he seemed like a good choice for a postdoc. Peter was turning out to be far more versatile and courageous than I expected. He had done a great job with our first Space Telescope data, but in Fall 1997, we weren't quite ready to say that the data required acceleration. In 1998, we had the evidence for acceleration. Now he was ready to take the next step. With Harvard graduate student Saurabh Jha, Peter looked into the problem. You needed to compute how well the two kinds of data agreed with every possible value of Ω_m and Ω_Λ. Saurabh made a plot that showed what you learned from the supernovae alone, and then what you gained from crossing it with the CMB data. I was stunned. When you combined the microwave background data with the supernova data, they made a bull's eye of probability. Those lines could have crossed anywhere, but the place they crossed was at $\Omega_m = 0.3$ and $\Omega_\Lambda = 0.7$. We had hit the treasure! This work was published in *The Astrophysical Journal Letters* in February 1998.[5]

The value for Ω_Λ wasn't, by itself, a powerful test—the fact that it was bigger than zero was the unique contribution of the supernova work. But the fact that the X came out at $\Omega_m = 0.3$ was something that we could compare to independent measurements of Ω_m, measurements that had nothing to do with supernovae or the microwave background. Following Zwicky's lead, there was a rich literature of measurements for the dark matter of the universe. Galaxy motions in clusters, gravitational lensing, and X-ray emission were all ways to detect invisible matter from its gravitational effects. And the measurements showed $\Omega_m = 0.3 \pm 0.1$. When completely independent lines of evidence converge, then you hear the ring of truth.

The story got even better at the beginning of 1999 when two balloon experiments designed to measure fluctuations in the microwave background reported their results. BOOMERANG, which had

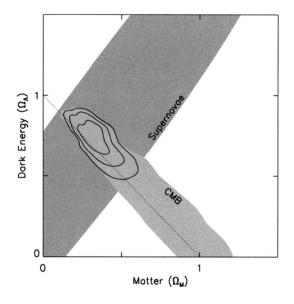

Figure 11.2. Combining information from supernovae and from fluctuations in the cosmic microwave background zeroes in on the values for Ω_m and Ω_Λ. Courtesy of Saurabh Jha; Harvard-Smithsonian Center for Astrophysics.

circumnavigated Antarctica and come back to its launching site after making 10 days of observations, and MAXIMA, another balloon experiment, both showed clean signals at the angular scale of 1°. What's more, the precision of these new measurements was enough to pin down the total Ω. They showed that $\Omega = 1.00 \pm 0.04$. Since $\Omega = 1.0000000000 \ldots$ is the value to which omega will be driven with exquisite accuracy by inflation, there are cheerful faces among the theorists. Inflation may not be the only model for the Big Bang, and there are variations on the inflation theme that do not produce $\Omega = 1$, but this was a test the purest version of this model could have failed. It did not fail and its several authors have good reason to be pleased.

While the cosmological pieces were rapidly falling into place like the cheerful frenzy of the last minutes of completing a jigsaw puzzle, Mike Turner, head cheerleader for cosmology, was writing

25°

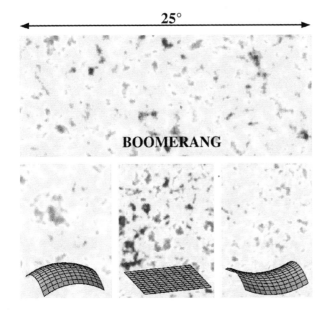

Figure 11.3. **Fluctuations in the microwave background from the BOOMERANG balloon experiment**. The measurements map the variations in the microwave background. The angular size of the fluctuations tells the cosmic geometry, which agrees best with a flat universe in which $\Omega_\Lambda + \Omega_m = 1$. Courtesy of the BOOMERANG collaboration. (Also see color insert)

articles. If you give a lot of talks at conferences, you may end up using similar material more than once. But a shift in the level of conviction that the data generate can be detected by careful reading of the titles for these talks. In March 1998, Mike Turner wrote one called "Cosmology Solved? Maybe." In April, he recycled the text with the less reserved title "Cosmology Solved?" And in October 1998, Mike took the next step when he entitled his talk, "Cosmology Solved? Quite Possibly." I'm looking forward to his future work, "Cosmology Solved!" Taking the cosmological constant out of Einstein's wastebasket seems to be required by the supernova data. Combined with the CMB measurements, the measurements point to a universe that is approximately two-thirds dark energy and one-third dark matter.

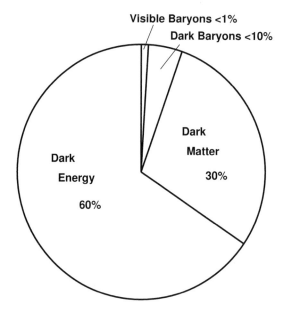

Visible Baryons <1%
Dark Baryons <10%

Dark
Matter
30%

Dark
Energy
60%

Contents of the Universe

Figure 11.4. **The Universal Pie**. Although we can be proud that we have filled up this diagram, the biggest slice of energy-density in the universe is dark energy, which we don't understand, and the next biggest is dark matter, which we don't understand. There is plenty of work to be done. Courtesy of Peter Garnavich, University of Notre Dame.

I guess we should be proud of the fact that we've been able to make any sense at all out of the universe, given our small brains, brief lives, and limited experiences, but there is something deeply unsettling about this picture. We may, quite possibly, have accounted for all of the matter and energy in the extravagant universe, but unfortunately, we don't know exactly what we are talking about. The dark energy could be the cosmological constant, but it could be something else that has negative pressure. And if inflation is right, then this is the second time the universe has been dominated by dark energy—once at 10^{-35} seconds, and now again at 10^{18} seconds after the Big Bang. The dark matter includes neutrons and protons and neutrinos, but most of it must be something else that is definitely not made of those familiar particles and is still unidentified. So, while we should take some pleasure in filling in the blanks, we've done it with things whose nature we only dimly grasp. This should be good news to somebody thinking of entering the field—the subject's not done, it is just starting.

Garnavich and Jha led the way for our team's modest foray into learning the nature of the dark energy. The cosmological constant will produce acceleration. We observed acceleration. This does not prove that the cosmological constant that Einstein imagined is responsible. What if there's something else that might cause acceleration? Could the dark energy be something else?

Looking at Einstein's equations, it is easy to see (remember I am not really a licensed theorist, but this is how they talk!) that you get acceleration from a component of the universe that has positive energy and negative pressure. While the cosmological constant does that, it has some other properties that are distasteful to theorists. First, the measured value is so small compared to theoretical estimates. They compute at least 10^{50}. We measure $\Omega_\Lambda = 0.7$. They prefer very large or exactly zero. But we measure something that is not very large and is not zero.

Second, the number we measure for dark energy, $\Omega_\Lambda = 0.7$, is not so different from $\Omega_m = 0.3$, the value for the dark matter. But that wasn't true in the past and it won't be true in the future if the dark energy is the cosmological constant. In the past, Ω_m was larger because the density was higher. We see evidence for that from the very high redshift supernovae, like SN 1997ff, where the data favor deceleration early in the history of the universe and acceleration only over the last 5 billion years. In the future, the energy density of dark matter will continue to fall, while a cosmological constant that stays constant would become a larger fraction of the total energy density in the universe. In other words, if there's a cosmological constant, it guarantees that Ω_Λ will eventually dominate Ω_m because the density of matter declines, while the energy density of the vacuum does not. Why do we live at the unique and cosmically brief moment when they are about the same?

You could say, "Well, that's just the way it is." But most theorists are, quite rightly, suspicious of coincidences. They don't like the smell of an idea that places us at a special time in the history of the universe. They would be happier if the dark energy were somehow related to the dark matter, so there would be a reason why they were so nearly the same. Paul Steinhardt, who pointed out that we might need Λ while the early LBL supernova results pointed the other way, has gone on to sketch a replacement for Λ he calls "quintessence." Quintessence is a form of vacuum energy that evolves with time, so the near agreement of the energy density in matter and energy is not a coincidence—it is just what you should expect. Paul has also gone on to challenge our imaginations and spelling ability with what he calls the Ekpyrotic universe, which replaces inflation with colliding space–time membranes at the beginning of time. The cosmological constant was invented by Einstein and it is completely consistent with general relativity as he formulated it. Quintessence and other forms of dark energy move beyond general relativity into new realms of physics. It is exciting to think that the ultimate origin of this effect, which is detected only by astronomical measurement over billions of light-years, is connected to the pursuit of understanding the universe on the smallest imaginable scales.

Is there a way to move the discussion of dark energy away from esthetics and rhetoric to measurement? Yes. We can distinguish some of the possibilities by measurement. The key ingredient is the way the pressure is related to the density. We call that the "equation of state." For ordinary gases like the carbon dioxide cartridge for a paint ball gun, as you increase the density, by stuffing more CO_2 molecules into the same-sized cylinder, the pressure goes up. For the cosmological constant, as the universe expands, the pressure (which is negative) and the energy density do not change. For other sources of acceleration, for example, quintessence, the pressure may change as the universe expands. This will leave a signature in the Hubble diagram if it makes cosmic acceleration occur at a different redshift. Garnavich showed from our early data that some forms of dark energy were ruled out, while the cosmological constant was completely consistent with the observations. But the present data are very sketchy. To do a good job on the cosmic equation of state, to find out what the dark energy really is, will require more numerous and more precise measurements of supernovae over a range of redshifts.

The Lawrence Berkeley Lab group has embarked on an ambitious plan to build a specialized satellite, deftly named SNAP (SuperNova Acceleration Probe) to discover and measure thousands of supernovae. This satellite would have a telescope with a wide field of view focused on an immense CCD array, far larger than anything ever sent into space on a civilian satellite. By concentrating on finding and measuring supernovae, SNAP advocates say they can pin down the nature of the dark energy. This is certainly worth doing, and I wish them well. One thing is for sure, they are going to go to a lot of boring meetings over the next ten years. I hope they know somebody down the hall who is working on projects with a shorter horizon.

But I suspect there will also be a role for less comprehensive, less expensive, and more rapid projects. You wouldn't want a quest for doing the perfect measurement to interfere with doing something that was pretty good. Perhaps by finding and measuring a few hundred supernovae from the ground we could learn the details of cosmic acceleration from redshift 0.5 up to redshift 1. This is the region where the effects of acceleration appear to be the largest, and we already know we can make the measurements from the ground, so detailed observations might pin down the equation of state for dark energy. Like the balloon-based experimenters measuring the CMB, we might find it possible to anticipate the main results of the SNAP satellite by a more modest route. That's part of our plan for the next few years. Maybe later we can check our results against the findings from SNAP.

While I've been sitting at my desk finishing this manuscript, a crew of astronauts has taken a bone-shaking ride on the Shuttle up to the Hubble Space Telescope. They skillfully installed a new camera with a larger field of view that takes sharper images. The new camera makes a practical proposition out of searching for distant supernovae with HST itself, as Stirling Colgate and Gustav Tammann prophesied 23 years ago. With this camera, a team led by Adam Riess hopes to find several objects beyond redshift 1, like SN 1997ff, to see if the universe really does have the stop-and-go property that signals the effect of dark energy at work. Even better, those deft astronauts plumbed in a new refrigerator for NICMOS, so, if all goes well, we will measure the light curves for these very distant supernovae in the infrared where cosmic expansion shifts their light. The next few years should be very exciting.

To see SN Ia even deeper into the cosmic past than we can observe with HST will require a telescope that has the power of the Space Telescope, but that is designed to work in the infrared. This telescope is already in development. The next generation space telescope will be a large, cold, space-based telescope that can see SN Ia (if there are any) back to redshift 5! Then we will certainly see how the ages of the stars in a galaxy affect the properties of supernovae, and we will be able to use supernovae to trace cosmic history right back to the very first stars. That's a telescope I've been willing to sit in meetings to help build.

What are the implications if this story is correct? If there is a cosmological constant causing acceleration over the last 5 billion years, then the universe will continue to accelerate indefinitely into the future. The expansion will literally be exponential: the bigger it gets, the more it speeds up. The universe will run away in headlong expansion. One curious effect is that galaxies we can observe today will get redshifted beyond our detection in the future. Instead of seeing more of the contents of the universe as time passes, we will see less and less. The universe could become a lonely, dull, cold, dark place. This is a good reason to do this work now. In a few hundred billion years, perhaps we won't be able to.

However, it is always unwise to think that today's best approximation to understanding the universe is really the whole story. The prevailing wisdom is always spoken in the same authoritative tone of voice, with the same degree of confidence. It's the content that changes. Ten years ago, a universe dominated by cold dark matter was very strongly advocated by many astrophysicists. This implied a "just right" universe with Ω_m equal to one that would expand and decelerate forever. Today, we say (in the same godlike tone of voice used to narrate planetarium shows and documentary films) that matter is only a fraction of the total energy density of the universe and that dark energy determines the future of cosmic expansion, which will accelerate forever. Since new evidence can change our best understanding on a timescale of ten years, we probably should be cautious in predicting what will happen in the next 100 billion years. If the present acceleration is caused by a variable sort of dark energy, it might go away at some distant time, ending the era of acceleration. And we shouldn't be too confident that there were not earlier episodes of acceleration that don't show up in the present data. The universe is wilder than we ordinarily dare to imagine. Although there is a sense today that the tumblers are all clicking into place as we unlock the secrets of the universe, this is not the end of the investigation, just the beginning.

It's a strange picture we have painted. The universe has dark energy and dark matter, neither of which is familiar to us from our everyday experience, or detected from any experiment on Earth. The visible part of the universe and the beautifully elaborate atoms that make up our bodies and our world are not the main material constituents of the cosmos. We have gone from thinking of ourselves as the centerpiece of creation though a series of cosmic leaps of understanding to seeing ourselves as observers and beneficiaries of a great pageant in space and time that we don't affect, but that has affected us greatly. We are not made of the type of particles that make up most of the matter in the universe, and we have no idea yet how to sense directly the dark energy that determines the fate of the universe. If Copernicus taught us the lesson that we are not at the center of things, our present picture of the universe rubs it in.

On the other hand, maybe the fact that we are not made of the stuff that forms most of the universe should make us feel special. Our origin is in the universe, with the atoms we're made of an unusual form of matter, baryons processed through stars. We're not the same as the dark matter or the dark energy, we're made of more versatile stuff that has more potential for complex, unpredictable outcomes like a human life.

A year ago, I was at a meeting of the American Physical Society, the biggest physics association in the United States. Up on the podium were a collection of presidential science advisers from the past, going all the way back to the Truman administration. Listening to them talk about the role of science in the United States, I grew irritated, impatient, and cross. They were talking about the value of science for economic growth through technical innovation. Science as the golden goose. They were talking about the value of science for national defense. Science replenishing secrets faster than they leak out. They were talking about the value of science to cure diseases and increase the span of human life. Science as the fountain of youth.

Now I suppose everybody wants to be rich, safe, and immortal. Or at least every Congressman. So I guess this is a reasonable set of goals for a president's science adviser to advocate. But the science of the universe is not aimed at creating wealth, improving defenses, or curing disease. It is aimed at increasing our understanding, and nobody on the platform was talking about that.

We have little brains and brief lives, but as I have tried to show in this book, we are beginning to build a rational picture of the universe in which we find ourselves. By combining the clues from ancient light and a hard-won understanding of how the world works, we are beginning to see the big picture for the history of the universe.

Is this important to people? Of course it is important to those of us who have the joy of working together to find things out. For us, it's an adventure. But is it important to others? I think so. People are curious and people have imagination. People want to know. Where did we come from? Where are we going? And when do we get there? Cosmology tries to answer those questions about the physical world using the best tools of modern technology and the best ideas that have been sifted and tested in the laboratory.

Part of the fun of cosmology is that it takes us into realms where laboratory physics doesn't yet reach. The accelerating universe is a new phenomenon, not discovered in a laboratory, which may open up a new area of physical understanding. It might lead to a new understanding of what the vacuum is, how gravity is related to the other forces, and how the contents of the universe shape its destiny. The elements of adventure, exploration, and the discovery of real things that are stranger than we dared imagine makes astronomy a well of inspiration, and not just for the experts. Our highest aspiration is not perfect comfort. If we were rich, safe, immortal, and *bored*, this would not be a vision of paradise. Cosmic discovery nourishes our deep desire to learn what the world is and how it works.

Epilogue

"We've done these calculations in a standard Λ–cold dark matter universe." The energetic young speaker at the front of the Phillips Auditorium at the Center for Astrophysics, Kathryn Johnson, a professor from Wesleyan, was setting the stage for presenting her new results on galaxy cannibalism. There were 100 people in the room for the Thursday Astronomy Colloquium, Kathryn had a lot of new results to share, and she wasn't wasting any of her time or theirs by justifying the cosmology she had assumed.

Nobody blinked. Nobody asked a question. But my mind, always unreliable after 4 p.m. in a darkened room, started immediately to drift into speculation. How could a "Λ" universe, two-thirds dark energy and one-third dark matter, be the "standard" picture in the autumn of 2003? Just 5 years earlier, cosmic acceleration had seemed unbelievable, and dark energy, in its guise as the cosmological constant, had been a notoriously bad idea, personally banished by Albert Einstein. What had changed?

Part of the answer is that the supernova results had gelled to become more solid. Further work by our high-z supernova search team had just come out in the September 1, 2003, issue of *The Astrophysical Journal*, and the Supernova Cosmology Project (SCP) had another paper in the works, too. Both teams still found that the universe was accelerating, 14 billion years after the Big Bang and 5 years after the first supernova results. We hadn't been fooled by bad luck or bad data—the new samples of type Ia supernovae at a distance of 5 billion light-years were, once again, 20% fainter than they would have been without cosmic acceleration. We could keep our goldfish, houses, and dogs.

And part of the answer was that some of the alternatives had been tracked down and shown *not* to be the cause of that slight dimming.

It probably wasn't pink pixie dust—this dust would have changed the colors of distant supernovae. We had measured those colors, at least for one supernova, and did *not* find any change. We have several more color measurements underway, but I am pretty sure we won't find a glaring problem with pixie dust.

And it probably was *not* the age of the stars. Both teams had used the Hubble Space Telescope (HST) to look at the galaxies in which distant supernovae were detected. On our team, Craig Hogan and Ben Williams had looked at the galaxies where supernovae were found. For the SCP, Richard Ellis and Mark Sullivan had inspected the sites of the explosions. When they sorted the host galaxies into elliptical galaxies (where there are very few young stars) and spirals (where there are many recently formed stars), they still got the same result for cosmic acceleration.

And it wasn't something that just depended on how long the light from a distant supernova took to get to us. Adam Riess had flipped his transparencies in public and shown that for one object at the highest redshift, SN 1997ff, the SN Ia was *not* fainter than you'd expect in a coasting universe, but brighter. This was the smoking gun of chapter 11 that showed *cosmology* was really responsible for what we see, not some error that grows larger as you look farther back. We live in a mixed dark-matter and dark-energy universe. At first, for about 7 billion years, the universe would slow down due to dark matter, then shift over to acceleration as expansion diluted the matter, while dark energy began to drive a more rapid cosmic expansion. The transition from slowing down to speeding up should be roughly 7 billion years in the past, halfway back to the Big Bang; SN 1997ff demonstrated that we could find and measure supernovae that far back, at least with the Hubble Space Telescope.

SN 1997ff was only one object, but in Hawaii, John Tonry and his student Brian Barris were leading the way to higher redshifts from the ground, using the big camera on the Subaru telescope to find a handful of supernovae out beyond redshift 1. The supernovae found in Hawaii also pointed in the direction of a stop-and-go universe. Adam Riess and the higher-z team were putting the new camera on the Hubble Space Telescope to good use, finding a dozen more objects that promised to confirm the earlier work.

Despite these improvements in the supernova data, I don't think that's the reason why our speaker spent no time in setting out the pros and cons of a Λ cosmology. The real reason was a sudden convergence of many independent lines of research on the very same values for the contents and age of the universe, weaving a web of evidence. Measurements of galaxy clustering from the Two Degree Field Galaxy Redshift Survey carried out in Australia were much more comprehensive than the Las Campanas Redshift Survey data shown in chapter 5, and the new data pointed to a universe with $\Omega_m \sim 0.3$. This was the same value we were getting from the supernova analysis.

The same was true with cosmic ages. Once you factored in the cosmic history prescribed by the supernova data of deceleration due to dark matter followed by acceleration due to dark energy, measurements of the Hubble constant from HST observations of cepheids (done just the way Hubble had done them in 1929 except in galaxies 25 times further away) gave an age for the universe of about 13.6 billion years. This was a very good match to the age of the oldest stars, as judged from the time it takes stars to fuse hydrogen into helium in their cores. The oldest globular clusters were recently gauged to be around 12.5 billion years old, allowing just enough time for them to form in the first 1 billion years after the Big Bang. When completely independent paths lead to the same place, it makes you think something good is happening.

All of this was very satisfying, but the most powerful new set of information came from better measurements of the cosmic microwave background (CMB). Since 1998, pioneering experiments had measured and reported the subtle texture of the exceptionally smooth glow from the Big Bang. Some used sensitive detectors carried high into the atmosphere on balloons, and others measured the CMB fluctuations from high, dry sites in the Atacama Desert of Chile and stations in Antarctica. Those preliminary results, when combined with the supernova measurements, matched well with a universe that had Ω_m of 0.3 and Ω_Λ of 0.7, as discussed in chapter 11. Like Los Angeles, the universe was one-third substance and two-thirds energy.

But the best was yet to come. In 2001, the Wilkinson Microwave Anisotropy Probe (WMAP), a satellite to measure the CMB, was launched into a unique orbit at 4 times the distance to the

Moon. From this superb perch, it patiently mapped the whole sky for a year.

The first data from WMAP were released in February 2003, and those results changed the tone of the discussion in observational cosmology from cautious and tentative to confident and quantitative. Just as the earlier measurements had indicated less precisely, the WMAP results confirmed that the universe has the large-scale geometry of flat space. This means that the 3-dimensional space we live in has none of the tricky curvature that Gauss imagined and Einstein showed how to compute. Just as expected, if inflation is right, the large-scale geometry of the universe is the geometry you learned in high school: parallel lines don't meet, the angles inside a triangle add up to 180 degrees, and the surface area of a sphere is $4\pi R^2$. Although general relativity provides for many other possibilities, the universe we live in is the simplest of these. When you combine the WMAP measurements with other evidence, the best age for the universe is about 13.4 ± 0.2 Gyr, and the best estimate for the present-day composition of the universe is $\Omega_m = 0.27$ and $\Omega_\Lambda = 0.73$.

These results are qualitatively the same as the earlier ones, so there was no big surprise when the WMAP results were presented (with vigorous tub-thumping from NASA) in February 2003, but rather an audible sigh of relief. Even though the CMB measurements don't detect cosmic acceleration directly, as the supernova measurements do, taken together, they point with good precision to a universe with both dark matter and dark energy. Things were fitting together—and the better you measured them, the better they fit. Quantitative agreement is the ring of truth. This is the reason why, by the autumn of 2003, our colloquium speaker didn't bother to make the case that a Λ-dominated universe was the right picture.

So, what's next? We are confident that dark energy is real, but *what is it?* Supernovae led the way in revealing cosmic acceleration. Can they now be used to pin down the nature of dark energy? We want to find out whether this weird stuff is really the cosmological constant Einstein created and discarded or possibly a more general "quintessence" that changes over time. If we could trace the onset of acceleration more precisely, we could tell if we were seeing something that is constant or something that is changing. But to do this, we need to improve our technique. More of the same isn't good enough. We need better precision.

Here's part of the plan: find 200 SN Ia in the next 5 years. Measure the light curve for each one in the same way and get a good spectrum of every one. If you do that, you should be able to tell whether dark energy has the same properties as the cosmological constant. Technically, we will find the "equation of state"—tracing the way the energy density of dark energy changes as the universe expands. This cannot fail to be interesting—either dark energy is the cosmological constant or it isn't. Either way, it is a deep mystery: there is still no explanation why the cosmological constant should be 10^{120} times smaller than the simplest theoretical estimates, and if dark energy turns out to be something else, that's also of tremendous interest. After all, whatever it is, it makes up two-thirds of the universe! We should take an interest.

To measure this property of dark energy, you need enough telescope time to find and follow 200 faint supernovae, enough computer power to sift through the data immediately before the supernova fades away, and a good acronym. We have all three. The ESSENCE (Equation of State: SupErNovae trace Cosmic Expansion, pronounced just like "SNs") program has been granted time on the 4-meter telescope at Cerro Tololo for 5-years worth of supernova hunting. We go to a small number of fields every 4 nights—that's frequent enough to get good light curves for the supernovae we discover, and the fields are big enough, thanks to the large CCD camera at the 4-meter, to make it very likely we will have some live supernovae in each month's series of observations. Chris Stubbs, recently at the University of Washington, but now at Harvard, has built a powerful dedicated computer system that can take each picture from an observing night, compare it to earlier images, and find all the new objects in our data by the next morning. We have time at the Gemini Observatories, VLT, Keck, and on the Magellan telescopes to get the spectra that will show us whether these are type Ia supernovae, and tell us whether the distant objects are the same as those nearby.

This enterprise, and similar work being carried out at the Canada–France–Hawaii Telescope, should build up a precise set of data for the era about 5 billion years ago, when the universe was switching over from deceleration due to dark matter to acceleration due to dark energy. The pace of the acceleration will tell whether this results from a dark energy that behaves like the cosmological constant or a dark energy that behaves differently as the universe expands. In a few years, we should have a grip on the nature of dark energy.

Similarly, the push to redshifts beyond 1 will show whether this picture is complete. The new "Advanced Camera for Surveys" on the Hubble Space Telescope is spectacularly good for searching out distant supernovae, measuring their light curves, and even at getting their spectra. In the next 2 years, we should go from a sample of 1 to a sample of 10 to a well-distributed sample of a few dozen high-redshift supernovae. These will show if we really do live in a stop-and-go universe with 7 billion years of cosmic deceleration from dark matter's drag followed by 7 billion years of cosmic acceleration powered by dark energy's push.

Technically speaking, the rate of change in an object's position is called velocity, the rate of change in velocity is called acceleration, and, to the unending amusement of physics students everywhere, the change in acceleration is called "jerk." So the search for the switch from cosmic deceleration by dark matter to cosmic acceleration by dark energy is a search for cosmic jerk. When Adam Riess gave a talk in October 2003 about preliminary results on the higher-z search that showed evidence for changes in cosmic acceleration, the headline writer at the *New York Times* couldn't resist running his picture under the banner: "A Cosmic Jerk that Reversed the Universe."

The Hubble Space Telescope is old enough now that we are beginning to think about the endgame for this splendid machine and the transition to the next big thing—the James Webb Space Telescope (JWST), a large orbiting telescope designed to function in the infrared, where the high-redshift supernovae shine. Earth's atmosphere is very thin at the altitude of 380 miles, where Hubble orbits, but there is a miniscule drag that, little by little, is lowering the orbit and, in time, will make HST spiral inward and enter the thick part of Earth's atmosphere, where it will burn up. NASA is planning to send the space shuttle up soon, once they are confident it will be safe to do this (perhaps in 2005), to install a new instrument on HST and boost the telescope to a higher orbit that should keep HST out of trouble for several more years.

NASA's original plan for the endgame was to use the space shuttle to go up to HST in 2012 (or whenever is the right time), put the telescope in the cargo bay, and bring it down to Earth to hang in the Air and Space Museum on the National Mall in Washington, D.C. After the *Columbia* tragedy in 2003, this no longer seemed like such a good idea. It's one thing to risk astronauts' lives to do something important that advances science—and science education is important—but it doesn't seem right to take that risk for a museum piece.

So now the discussion centers on ways to extend the life of HST and to bring the 12-ton satellite down in a controlled way when the time comes. We want to be sure it eventually ends up in the Indian Ocean, not in Indianapolis. NASA engineers are working hard to design a small rocket that could be attached to HST during a shuttle servicing mission in 2010. The shuttle could boost HST for operation until JWST is ready, and then that small rocket could be used to bring down HST in a controlled way, sometime after 2012.

As a scientist, I think it would be a good thing—if NASA *must* take the risk and bear the expense of making a shuttle trip to install that bring-it-down rocket in 2010—to use that final servicing mission to install an even better camera that could be used for the dark-energy problem. A high-powered committee, headed by John Bahcall, has recommended that NASA consider how to use that shuttle trip to get more science from HST, if possible. We will see how all of this turns out. We need to balance the desire to get on with JWST against continued operation of the existing space telescope. One possibility would be to simplify the operations of HST in its final years by concentrating on the dark-energy problem with a wide-field camera taken up on the final servicing mission.

Meanwhile, the Lawrence Berkeley Lab, home of the Supernova Cosmology Project, is investigating the possibilities for SNAP (SuperNova Acceleration Probe), a satellite dedicated to the study of dark energy. It is a very ambitious program, with a proposed telescope almost as big as HST, a billion-pixel camera with hundreds of CCDs, and a spectrograph envisioned to work superbly from ultraviolet to infrared wavelengths. If SNAP is built and performs as specified, it would provide a large and uniform set of 2000 well-observed supernovae that should narrow down the nature of dark energy to a few percent by some time around 2014. Like all large satellite programs, the SNAP team will require persistence, fortitude, good luck, and a boatload of money to reach its goals. I wish them well.

And finally, we turn to the realm of theory and experiment on the very small scale. As a (slightly fraudulent) member of the International Brotherhood of Theorists (through my experience at the Institute for Theoretical Physics), I hope that there will be a theoretical breakthrough to match the progress in observations. There are now 1,545 papers on the High Energy Physics preprint server that cite our original evidence for an accelerating universe. Most of them are theory papers. I can't claim to have read them all. But, as I understand it, there's no great progress to report on answering the question, Why is the vacuum energy so much smaller than predicted from a simple calculation? Although it may be possible to make a string theory of all the forces that reside in 11 dimensions, and gravity may seep into those extra dimensions, we don't yet know why gravity would make the cosmic acceleration small but real. One possible path might come from the experimental world of particle accelerators, where investigators hope to see signs of "supersymmetry," a theory of particle physics that goes beyond today's Standard Model. In supersymmetry, the lightest particle, dubbed the "neutralino," might be a good candidate for the dark matter, and a cancellation of sorts might be able to make the dark energy small. Of course, this discussion would be more convincing if experiments at Fermilab or at CERN show that the neutralino exists in the real world as well as in the minds of theoretical physicists.

But it is certainly possible that a conceptual advance will show us why the cosmological constant is 0.7. The present quantitative disagreement is so large it would count as a great success if the prediction were 7 or 70 or 700 in those units. We shall see. Cosmology is on the minds of the string theorists, and those are very active minds.

The discovery of cosmic acceleration has been a tremendous adventure in finding out how the world works. Although we bravely claim to be entering the era of "precision cosmology," there are still great voids in our understanding. While we are beginning to determine the amounts of dark energy and dark matter to a few percent, we still don't know what either of them is. But we do know what to do next, and we are eager to get on with the hunt. This story is not over. In fact, the fun has just begun.

notes

Useful References

Related information is available at http://cfa-www.harvard.edu/~rkirshner.

Easy access to the original astronomical literature is available through the Astrophysical Data System: *http://adswww.harvard.edu/*

Donald Goldsmith, *The Runaway Universe*, Perseus Books, Cambridge, MA, 2000. A vigorous account of this work by an experienced science writer.

Tom Lucas, *Runaway Universe*, a 1-hour NOVA documentary available from *http://main.wgbh.org/wgbh/shop/wg2713.html*

Mario Livio, *The Accelerating Universe*, Wiley, New York, 2000. An interesting esthetic approach to science and to the implications of this work.

Ken Croswell, *The Universe at Midnight*, Free Press, New York, 2001. An excellent survey of modern astrophysics.

Lawrence Krauss, *Quintessence*, Basic Books, New York, 2000. A lively sketch of the microscopic physics of dark matter and dark energy.

Alan H. Guth, *The Inflationary Universe*, Addison–Wesley, Reading, MA, 2000. A first-hand account of the inflationary universe.

Martin Rees, *Our Cosmic Habitat*, Princeton University Press, Princeton, N. J., 2001. Adventurous thoughts on cosmology by one of the experts.

Laurence Marschall, *The Supernova Story*, Princeton University Press, Princeton, NJ, 1994. Breezy and readable account of SN 1987A and other supernovae.

Preface

1. In fact, navigation books caution strongly *against* judging the distance of a light by its apparent brightness. The penalty for error in coastwise navigation is more severe than in cosmology.

2. As quoted in the superb biography of Einstein by Abraham Pais, *Subtle Is the Lord*, p. 288, Oxford University Press, Oxford, UK, 1982.

Chapter 1: The Big Picture

1. A person wading on the shore whose eye is 1.6 meters (5 feet 4 inches) above sea level can see a distance of about 10 kilometers (6 miles) to the horizon. Even though curvature is the cause of a finite distance to

the horizon, most of us have no common-sense feeling for living on a curved planet. This may be because haze is important at this distance and it makes objects at a distance of 10 kilometers sometimes visible and sometimes not. If the Earth were much smaller than its radius of 64 million meters, like a small asteroid, the horizon would be closer and we might know we lived on a curved surface.

2. In 1801, Gauss, then 24, used early and incomplete observations of the first asteroid, Ceres, to predict where it would reappear when it emerged from the glare of the sun. His prediction, based on Newtonian gravity, was right on the money, and the new object was recovered on 31 December 1801. Predicting the orbit of Ceres was the beginning of an auspicious public career for Gauss as an astronomer.

3. A foot, 0.3048 meter, is a unit of length used in the United States, Liberia, and Myanmar. In this book, when I say a billion or a billionth, I mean 10^9 or $1/10^9$.

4. Time machines are brilliantly, amusingly, and seriously discussed in J. R. Gott's book *Time Travel in Einstein's Universe*, Houghton Mifflin, New York, 2001.

5. The speed of light is 2.997929×10^8 meters per second and a year is about 3.155×10^7 seconds. So a light-year is about 9.46×10^{15} meters.

Chapter 2: Violent Agents of Cosmic Change

1. Comte is quoted in *Inward Bound*, Oxford University Press, Oxford, UK, 1988, by Abraham Pais, p. 165.

2. The information contained in the absence of light is analogous to the curious incident of the dog in the night-time. In the story *Silver Blaze*, Inspector Gregory asks Sherlock Holmes,

"Is there any point to which you would wish to draw my attention?"
"To the curious incident of the dog in the night-time."
"The dog did nothing in the night-time."
"That was the curious incident," remarked Sherlock Holmes.

The notion that the methods of Sherlock Holmes provide a template for discovering modern physics has been explored to its outermost limit by *The Einstein Paradox* by Colin Bruce, Perseus Books, Reading, MA, 1997.

3. A detailed account of the biblical and physical methods for learning the age of the universe is given in *Measuring Eternity* by Martin Gorst, Broadway Books, New York, 2001. Gorst is much better on the subjects of Bishop Ussher and Lord Kelvin than on the ages of stars and the facts of discovering the accelerating universe.

4. Fred Hoyle wrote *Frontiers of Astronomy*, which I remember reading as vividly as any Sherlock Holmes story. Fred died in 2001. Hoyle's autobi-

ography *Home Is Where the Wind Blows* (University Science Books, Mill Valley, CA, 1994) tells how Hoyle's interest in astrophysics was stirred by his encounter with Walter Baade on an unauthorized visit to Pasadena during World War II, while Baade was restricted to Pasadena and Mount Wilson as an enemy alien. Baade's vivid account of the physical problems posed by supernovae drew Hoyle into the field.

5. The phenomenon of gravitational waves and the attempts to detect them is described in the sprightly book by Marcia Bartusiak, *Einstein's Unfinished Symphony*, Joseph Henry Press, Washington, D.C., 2000.

6. The publication "Evidence for $^{56}Ni \rightarrow {}^{56}Co \rightarrow {}^{56}Fe$ Decay in Type Ia Supernovae" appeared in *The Astrophysical Journal Letters* **426**, L89 (1994). It concludes, "This is a simple, direct, and satisfying (if not iron-clad) demonstration of the ^{56}Ni decay model for SN Ia."

7. Tycho's account is quoted in *The Historical Supernovae* by David H. Clark and F. Richard Stephenson, p. 174, Pergamon Press, Oxford, U.K., 1977.

Chapter 3: Another Way to Explode

1. These are the temperature units named for Darwin's nemesis, Lord Kelvin: they have their origin at absolute zero ($-273°$ centigrade, $-460°F$) and are the same size as centigrade degrees. Water boils at $+373$ kelvins.

2. W. Baade and F. Zwicky, *Proceedings of the National Academy of Sciences (U.S.A.)*, **20**, 254 (1934).

3. Additional variations on the type I theme are elaborated in Chapter 8. The original type I of Zwicky and Baade is now called type Ia.

4. Supernovae are given alphabetical labels in order of the reports to the Central Bureau for Astronomical Telegrams (now mostly e-mail) of the International Astronomical Union. So the first supernova of 1987 was SN 1987A, the second SN 1987B. When we get to the end of the alphabet, we shift to two letter designations: aa, ab, ac,. . . . In 2001, as I write this, the last supernova of the year was 2001it, which means there were a total of 254 discovered. (That's 26 with single letters plus 8 more double letters with 26 supernovae each from "aa" through "hz," plus 20 more from "ia" to SN 2001it.)

5. A lavishly illustrated account with photographs by Roger Ressmeyer of the SN 1987A observations is my article "Supernova—Death of a Star" in *National Geographic* **173**, 618 (1988). But if you want the schadenfreude of my flirtation with error concerning Sanduleak-69 202, this was vividly chronicled by Robin Bates in his NOVA documentary, "Death of a Star." The video is available from WGBH in Boston at *http://main.wgbh. org/wgbh/shop/wg1411.html*

An excellent popular-level survey of supernovae, and especially of SN 1987A is Lawrence Marschall's *The Supernova Story*, Princeton University Press, Princeton, NJ. 1994.

6. "Submillisecond optical pulsar in supernova 1987A" by Kristian, Pennypacker, Middleditch, Hamuy, Imamura, Kunkel, Lucino, Morris, Muller, Perlmutter, Rawlings, Sasseen, Shelton, Steinman-Cameron, & Tuohy, *Nature* **338**, 234 (1989).

7. John Middleditch has published a further analysis of the pulsed emission from SN 1987A in *New Astronomy* **5**, 243 (2000). The standard of proof should be higher the second time around. There is, as yet, no independent confirmation of this work, so it seems prudent to reserve judgment.

Chapter 4: Einstein Adds a Constant

1. This direct quotation is from Abraham Pais, *Subtle Is the Lord*, p. 257, Oxford University Press, Oxford, U.K. 1982.

2. An arcsecond is an angle: 1/3600 of a degree. It is well below the threshold of acuity for human vision, which is closer to 60 arcseconds (1 arcminute). A poppyseed 1 mm across at a distance of 200 meters covers 1 arcsecond.

3. Pais, *Subtle Is the Lord*, p. 253.

4. Quoted in the touching sketch of Eddington's life written by Subramanyan Chandrasekhar, *Eddington: The Most Distinguished Astrophysicist of His Time*, Cambridge University Press, Cambridge, UK, 1983.

5. This well-known story is recounted briefly in Pais, pp. 304–312, in more detail in Chandrasekhar's slim book, and in three-part harmony with unusual attention to Freundlich's 1914 Crimean expedition to measure the deflection that was cut short by the outbreak of World War I as told by Amir C. Aczel in *God's Equation*, Delta Books, New York, 1999. Aczel is not quite so comprehensive or accurate in his account of the accelerating universe.

6. Einstein originally published his ideas on cosmology in the Prussian Academy of Sciences Session Reports, p.142 (1917).

Chapter 5: Cosmic Expansion

1. A Tucson developer threatened to sue me for $900,000,000 when I wrote to a zoning hearing concerning the deleterious effects of a proposed large housing development on our astronomical observatory. The Whipple Observatory at Mount Hopkins is the place where many of our supernova data have been obtained. Fortunately, I understand powers-of-ten quite well, and felt no fear that I would ever have to pay such a sum.

$900,000 or even $90,000 would have been more intimidating. The zoning change was not granted.

2. A. S. Eddington, *The Mathematical Theory of Relativity*, p. 162, Chelsea, New York, 1975.

3. H. S. Leavitt in *Annals of the Harvard College Observatory* **60**, 97–108 (1908).

4. An interesting modern account of the discussion between Curtis and Shapley is given by Virginia Trimble in *Publications of the Astronomical Society of the Pacific* **107**, 1133 (1995).

5. E. P. Hubble, *The Astrophysical Journal* **69**, 103 (1929)

6. Although the units of the Hubble constant are a bit mixed, they are suitable for the subject. Redshifts can be expressed as velocities in kilometers per second and distances in parsecs (1 parsec = 3.262 light-years = 3.086×10^{16} meters) or megaparsecs (millions of parsecs). A parsec is a good unit, because the distances to nearby stars are a few parsecs and megaparsecs are good units because the distances to nearby galaxies are a few megaparsecs. A similar unit that is not standard, but useful in a particular context is the Smoot, used at MIT. A Smoot is the length of Oliver R. Smoot 1962: 5 feet 7 inches. The length of the Harvard Bridge, across which MIT undergraduates must walk on freezing Massachusetts nights is 364.4 Smoots plus one ear.

7. The Boötes Void was reported in *The Astrophysical Journal Letters*, **248**, L47 (1981) by R. P. Kirshner, A. Oemler, P. Schechter, and S. A. Shectman.

8. This survey pioneered the large-scale use of electronic cameras scanning across patches of the sky to select the galaxies and the use of fiber optics to obtain the spectra of many galaxies simultaneously. The Las Campanas Redshift Survey is described in S. A. Shectman, S. D. Landy, A. Oemler, D. L. Tucker, H. Lin, R. P. Kirshner, and Paul L. Schechter, *The Astrophysical Journal*, **470**, 172 (1996).

9. George Gamow, *My World Line*, Viking Press, New York, 1970. Gamow was a wonderfully creative and playful person, but perhaps not the world's most reliable narrator. Still, the quote is too good to resist. Print the legend.

10. A. S. Eddington, *The Expanding Universe*, Cambridge University Press, Cambridge, UK, 1987. This book is, as noted in the introduction by Sir William McCrea, "a maddening production," with the cosmological constant as its main subject. "The reader can never be sure when he is being invited to follow a serious arguments, or when he is being—oh so delicately—conned!"

11. Eddington, *The Expanding Universe*, p. 102.

12. The detailed comparison of general relativity with the evidence is laid out for the general reader in Clifford Will's excellent book, *Was Einstein Right?* Basic Books, New York, 1993.

13. A. Einstein, and W. de Sitter, *Proceedings of the National Academy of Sciences* **18**, 213 (1932).

14. Chandrasekhar's book on Eddington cited above tells this curious anecdote, which seems too neatly symmetric to be the whole story of Einstein, de Sitter, and Λ.

15. The critical density is given by $3H_o^2/8\pi G$, where G is Newton's gravitational constant. For $H_o = 70$ kilometers per second per megaparsec, this is 1×10^{-26} kilogram per cubic meter—a very low density, indeed! Then Ω is the actual density divided by the critical density, so it is exactly one when the observed density is equal to the critical density.

16. A. R. Sandage, "The Ability of the 200-Inch Telescope to Discriminate Between Selected World Models," *The Astrophysical Journal* **133**, 355 (1961).

17. A universe described by just two numbers is too simple. For a modern view that includes the fluctuations from the Big Bang and their effect on the growth of structure in the universe, see *Just Six Numbers* by Martin Rees (Basic Books, New York, 2000).

18. For a recent scientific portrait of Sandage, Dennis Overbye's *Lonely Hearts of the Cosmos*, is unique. Be certain to get the 1999 paperback edition (Little, Brown, Boston) with the precisely observed Afterword on cosmic acceleration.

Chapter 6: What Time Is It?

1. Martin Gorst, *Measuring Eternity*, Broadway Books, New York 2001.

2. To compute the Hubble time, we need to get all the quantities in the same units. $t_o = 1/H_o$, but H_o is in units of kilometers per second per megaparsec.

Since 1 kilometer = 10^3 meters and 1 megaparsec = 3.08×10^{22} meters, then a Hubble constant of 70 kilometers per second per megaparsec = 70×10^3 meter/kilometer $\times 1/(3 \times 10^{22})$ meter/megaparsec = 2.27×10^{-18} second^{-1}.

Strictly speaking, this is the correct way to express the Hubble constant. It means that the local patch of the universe stretches out by 2.27×10^{-18} of its present size each second. Once the units are straightened out, the computation of the Hubble time is simple: $t_o = 1/(2.27 \times 10^{-18}$ second$^{-1}) = 4.40 \times 10^{17}$ seconds. So that's the age of the Universe, based on the present expansion rate. Since a year has 3.16×10^7 seconds, this can be expressed as $t_o = 4.40 \times 10^{17}$ seconds$/3.16 \times 10^7$ seconds/year = 13.9×10^9 years. Fourteen billion is close enough. Since we're not completely sure this is the final value for the Hubble constant, for some other value of H_o you would get t_o = 13.9 billion years × (70 kilometers/second/megaparsec / H_o) This "Hubble time" would be the age of the universe if the rate of expansion does not change. Gravity from dark matter slows expansion and dark energy accelerates it.

3. The legend is that Eddington was approached by a reporter who asked whether it was true that there were only three people in the world who understood general relativity. Eddington didn't answer. "Come, come, sir, don't be modest."

"I was just trying to think who the third might be."

4. A. S. Sharov and I. D. Novikov, *Edwin Hubble, the Discoverer of the Big Bang Universe* p. 67, Cambridge University Press, Cambridge, UK, 1993.

5. A. S. Eddington, *The Expanding Universe*, p. 65.

6. Amazingly, Hans Bethe is still making contributions to astrophysics: he has been working effectively on the mechanisms of supernova explosions and published a paper on that topic in 2001.

7. In 1999, confusion between English units and metric units led to the Climate Orbiter burning up in the Martian atmosphere.

8. An amusing chart compiled by my colleague John Huchra shows the quoted values of the Hubble constant from 1929 to the present: *http://cfa-www.harvard.edu/~huchra*

9. N. Panagia, R. Gilmozzi, F. Macchetto, H.-M. Adorf, and R. P. Kirshner, "Properties of the SN 1987A Circumstellar Ring and the Distance to the Large Magellanic Cloud," *The Astrophysical Journal Letters* **380**, L23 (1991).

10. R. G. Eastman, and R. P. Kirshner, "Model Atmospheres for SN 1987A and the Distance to the Large Magellanic Cloud," *The Astrophysical Journal* **347**, 771 (1989).

Chapter 7: A Hot Day in Holmdel

1. Astronomers are familiar with electrons being ripped off hydrogen atoms by ultraviolet photons near hot stars, and then emitting visible light as the atoms reassemble themselves. In the trade, this is called "recombination." The light from recombining hydrogen atoms makes gas clouds in star-forming regions glow. It isn't quite logical, but we also call the era when hydrogen atoms first formed in the cooling aftermath of the Big Bang, "recombination." It isn't logical because it isn't "re" combination when it's the first time. But that's we call it. It would be couth to call it combination, but our advertant speech is maculate.

2. When we were in Munich not long ago, Jayne and I wandered through the endless halls of the Deutches Museum—a giant and thorough

museum of science and technology. There were zeppelin parts and full-sized ships, a spark-crackling van de Graff generator, and didactic exhibits on electromagnetism that would take about a year to assimilate. But up in the astronomy hall, I was astonished to see the original receiver that Penzias and Wilson had used. There, on its chart recorder, was the actual signature of the hot Big Bang. What was it doing in Munich? Penzias, born in Munich in 1933, moved with his family to the United States in 1940. I guess he had good memories of the Deutches Museum. Perhaps that is where he learned about electromagnetism!

3. This is not the whole story. The motion of the Milky Way through the photons of the microwave background does create a fore-and-aft effect of about 2 parts in 100. Once this simple motion is removed, the small-scale roughness is of order 1 part in 100,000.

4. Guth kept a diary, which was open to this page in an exhibit at the Adler Planetarium in Chicago. All of this is recounted in his book *The Inflationary Universe*, Helix Books, Reading, MA, 1997.

5. There are versions of inflation that lead to a universe where Ω is not exactly one. Some of this is described in chapter 4 of J. Richard Gott's brilliantly hued *Time Travel in Einstein's Universe*, Houghton Mifflin, Boston, 2001.

6. Firsthand accounts are in George Smoot and Keay Davidson's *Wrinkles in Time*, Avon Books, New York, 1993, and *The Very First Light* by John Mather and John Boslough, New York, Basic Books, 1996.

7. The temperature of Hell is reported by Dante and others to be the temperature at which brimstone (sulfur) melts. This is 718 kelvins. So we're talking about the universe when it was 60 million times hotter than Hell.

8. This subject is elegantly discussed by Steven Weinberg in his classic book, *The First Three Minutes*, Basic Books, New York, 1993.

9. Helium in birthday balloons is not exactly straight from the Big Bang. It comes from the radioactive decay in the Earth of more complex elements that formed in stars. The world's biggest source of helium is a natural gas field near Amarillo, Texas. At the Helium Monument, helium's indifference to combining chemically with other atoms is illustrated by a piece of apple pie in a container filled with helium. It looks as fresh as the day it was baked by the infrared photons of an oven. It is 33 years old (*www.dhdc.org/heliummonument.htm*).

10. Zwicky's initial work was published in *Helvetica Physica Acta* **6**, 110 (1933). His book *Morphological Astronomy*, SpringerVerlag, Berlin, 1957, contains a longer English version. A recent review by Sidney van den Bergh in *Publications of the Astronomical Society of the Pacific* **111**, 657 (1999) is very entertaining.

Chapter 8: Learning to Swim

1. R. Minkowski, *The Astrophysical Journal* **89**, 156 (1939).

2. R. Minkowski, *Publications of the Astronomical Society of the Pacific* **52**, 206 (1940).

3. The details were kindly provided by Robert N. Turner, Assistant Curator at the New Mexico Museum of Space History, Alamagordo, New Mexico.

4. R. P. Kirshner, and J. Kwan, *The Astrophysical Journal* **193**, 27 (1974).

Chapter 9: Getting It First

1. Planetary nebulae have nothing to do with planets. All elements beyond helium are called "metals." The apparent brightness of stars is measured on a descending scale of "magnitudes" that has a standard interval of $(100)^{1/5}$. There are many other examples of historical fragments that baffle outsiders.

2. S. A. Colgate, *The Astrophysical Journal* **232**, 404 (1979) and G. A. Tammann, IAU Symposium #101, *Scientific Uses of the Space Telescope.*

3. If there are something like 10^{10} galaxies in the observable universe, then since a century is 3×10^9 seconds, a supernova per century per galaxy means there are about 3 supernovae every second in the universe. The problem isn't a shortage of supernovae, it's that we can't look far enough in all directions.

4. The "News & Views" article entitled "Explosive Assault on Ω" appeared in *Nature* **339**, 512 (1989).

5. Muller's work on comet crashes, exploding stars, and the hot Big Bang are described in his biographical book written with Philip M. Dauber, *The Three Big Bangs*, Helix Books, Reading, MA, 1996.

6. This scientific detective story is vividly told in *T. rex and the Crater of Doom* by Walter Alverez, Princeton University Press, Princeton, NJ, 1997.

7. This appeared in a long article by John Noble Wilford in the *New York Times* in April 1998 quoted in Wilford's book *Cosmic Dispatches*, p. 248, W.W. Norton, New York, 2001.

8. The printed version of the proceedings is published as Neil Turok (editor), *Critical Dialogs in Cosmology*, World Scientific Publishing, Singapore, 1997. It doesn't contain the debates, though there is an article by Fukugita that explores the supernova-based case against the cosmological constant. Curiously, two members of the high-z supernova team went to high school in East Brunswick, New Jersey: Saurabh Jha and David Reiss. Adam Riess grew up in Warren, New Jersey. John Tonry went to Princeton. I myself was born in Long Branch. Is it the water or the microwave background?

Chapter 10: Getting It Right

1. S. Perlmutter et al. *The Astrophysical Journal* 483, 565 (1997).

2. Alex Filippenko has written a racy first-hand account of the events in this chapter for the *Publications of the Astronomical Society of the Pacific* **113**, 1441 (2001).

3. In Chile, this is the polite form of "no."

4. The slowing of cosmic clocks manifested by stretched-out supernova light curves was proposed in 1939 by Olin Wilson of the Mount Wilson staff as evidence that the cosmic redshift was really caused by expansion and not by the "gradual dissipation of photon energy," the "tired light" hypothesis Zwicky proposed in 1929. (O. C. Wilson, *Astrophysical Journal* **90**, 634 (1939)). This effect was sought by Bert Rust in his Ph.D. thesis in 1974, but the data were not adequate. Evidence for time dilation in SN 1995K was published by Bruno Leibundgut and the high-z team in *Astrophysical Journal Letters* **466**, L21 (1996). Gerson Goldhaber and colleagues published a similar conclusion in a conference report in *Thermonuclear Supernovae*, edited by Pilar Ruiz-Lapuente and J. Isern, Kluwer, Dordrecht, The Netherlands, 1997, p. 777. A more thorough analysis by Goldhaber is in the *The Astrophysical Journal* **555**, 359 (2001).

5. This work was submitted to the *Astrophysical Journal Letters* on 14 October 1997 and published in the 1 February 1998 issue as P. M. Garnavich, et al., *Astrophysical Journal Letters* **493**, L53 (1998).

6. As of this writing, this talk is still available at the Institute for Theoretical Physics website: *http://online.itp.ucsb.edu/online/plecture/kirshner/*

7. In fact, Sean Carroll was so bright and interesting he got most of his thesis advice 3000 Smoots away at MIT! The article on the cosmological constant appeared in *Annual Reviews of Astronomy and Astrophysics* **30**, 499 (1992). Sean is now on the faculty at the University of Chicago.

8. The units of Λ and Ω_Λ are connected by $\Omega_\Lambda = \Lambda c^2 / 3H_o^2$.

9. One of the people down the hall from me at the ITP was Tony Zee, author of my favorite popular-level book on gravity, *An Old Man's Toy*, Collier Books, New York, 1989. As he puts it, "The discrepancy is so large that no amount of squirming could get {the physicists} off the hook."

10. There are a number of accounts of this event. The "Afterword," pp. 426–436, to Dennis Overbye's *Lonely Hearts of the Cosmos*, cited above, describes these events very precisely. Similarly, Ted Anton's *Bold Science*, W. H. Freeman, New York, 2000, contains a carefully observed profile of Saul Perlmutter's team. It describes his team's remorse at not being more definite about the evidence for cosmic acceleration in January 1998. Jim Glanz's article appeared in *Science* **279**, 651(1998), one of many he wrote that year. Glanz describes his source for that story in *Astronomy* (October 1999), p. 94.

11. This conference report to the 3rd International Symposium on Sources and Detection of Dark Matter appears as A. V. Fillipenko and A. G. Riess, "Results from the High-z Supernova Search Team," *Physics Reports* **307**, 31–44 (1998).

12. Goldhaber is quoted in *Cosmic Dispatches*, p. 247.

13. The Höflich, Wheeler, and Thielemann paper is in *The Astrophysical Journal* **495**, 617 (1998).

14. Alison Coil's paper with the high-z team, "Optical Spectra of Type Ia Supernovae at $Z = 0.46$ and $Z = 1.2$," appeared in *The Astrophysical Journal* **544**, 111 (2000).

Chapter 11: The Smoking Gun?

1. The last time I looked, there were 1,545 citations in the physics and astrophysics literature to the high-z team's Riess et al. 1998 paper in *The Astronomical Journal*. A typical number for an astronomy paper is somewhere around 20 citations, often in later papers by the same authors! These papers and preprints can be found through the preprint server: *http://arxiv.org*

2. Tomasso Giovanni Albinoni (1671–1751), born in Venice, wrote 48 operas (most have been lost) and was one of the first to write concertos for solo violin. I had to look this up. I was familiar with Dudley Do-Right from a misspent childhood watching *Rocky and Bullwinkle*.

3. John's conference proceedings are available at the Los Alamos preprint server: *http://xxx.lanl.gov/abs/astro-ph/0105413*

4. No culture that has invented the codex (edge-bound books) has ever gone back to the scroll. Yet Powerpoint presentations are like the items on a scroll, with an order that is fixed and hard to scan. But I digress.

5. P. Garnavich, and the high-z team, "Constraints on Cosmological Models from Hubble Space Telescope Observations of High-z Supernovae," *Astrophysical Journal Letters* **493**, L53 (1998).

acknowledgments

It takes many people to reveal an accelerating universe, and I am extremely grateful to all my high-z colleagues for letting me share in this adventure. I have discovered that it also takes quite a few people to finish a book, and I am grateful for the help I have received. First, I must thank my wife, Jayne Loader, for her encouragement, apt suggestions, and sharp editing pencil. She has proved to be a nimble Web-ferret—putting her finger promptly on needed facts. Jack Repcheck, late of Princeton University Press and now at Norton, got me going on this book, and Joe Wisnovsky, late of Norton and now at Princeton, got me to the finish line. But I would not have started this or any other book without accumulating a deep debt to my parents, Dick and Virginia Kirshner, who, among other things, let me build Tesla coils and oscilloscopes, apparently without fearing my imminent electrocution. My sister, Sarah Kirshner, helped more directly, by expertly reading an early draft and helping me haul away debris to reveal a book within. My daughter, Rebecca Rand Kirshner, drove a stake through the heart of the opening screenplay, transmogrifying it into a Hawaiian dream. Matthew Kirshner's drawings were invaluable.

My recent graduate students on this work, Chris Smith, Brian Schmidt, Adam Riess, and Saurabh Jha, have been a joy, and my postdocs Bruno Leibundgut, Peter Garnavich, and Tom Matheson have been outstanding. Pete Challis brings relentless good cheer and energy to this project, and was especially helpful in preparing figures for this book. The National Science Foundation has steadily supported supernova research at the Harvard–Smithsonian Center for Astrophysics and we have also received generous support from NASA through the Space Telescope Science Institute.

My high-z teammates have been tolerant and helpful in getting this manuscript whipped into shape. Though they helped me remove many errors, any that remain are mine. Alex Filippenko read the manuscript with a microscope, Adam Riess found his old e-mail,

Brian Schmidt helped straighten out my memory, Bruno Leibund-gut explained the equivalence of *mass* and energy, and Saurabh Jha helped me get the large-scale structure right. Our high-z colleagues in Chile, including Nick Suntzeff, Mario Hamuy, Bob Schommer, and Mark Phillips, have made major contributions to supernova studies and to this work in particular. I have tried to tell their story, too. It is tragic that Bob, who spent many nights at Cerro Tololo finding high-z supernovae, died in 2001 and I have no way to thank him for his friendship and his zeal for this work. We all miss him.

index